CHEMISTRY DIVISION
General Editor: SIR ROBERT ROBINSON

BASIC TECHNIQUES OF PREPARATIVE ORGANIC CHEMISTRY

BASIC TECHNIQUES OF PREPARATIVE ORGANIC CHEMISTRY

by

WILLIAM SABEL, B.Sc., F.R.I.C.
Principal Lecturer in Industrial Chemistry,
Oxford College of Technology

PERGAMON PRESS
OXFORD · LONDON · EDINBURGH · NEW YORK
TORONTO · SYDNEY · PARIS · BRAUNSCHWEIG

Pergamon Press Ltd., Headington Hill Hall, Oxford
4 & 5 Fitzroy Square, London W.1
Pergamon Press (Scotland) Ltd., 2 & 3 Teviot Place, Edinburgh 1
Pergamon Press Inc., 44–01 21st Street, Long Island City, New York 11101
Pergamon of Canada, Ltd., 6 Adelaide Street East, Toronto, Ontario
Pergamon Press (Aust.) Pty. Ltd., Rushcutters Bay,
Sydney, New South Wales
Pergamon Press S.A.R.L., 24 rue des Écoles, Paris 5e
Vieweg & Sohn GmbH, Burgplatz 1, Braunschweig

First edition 1967
Library of Congress Catalog Card No. 67–24315

Printed in Great Britain by A. Wheaton & Co., Exeter and London

(3209/67)

CONTENTS

PREFACE

ORGANIC chemistry is still an experimental science, and the study of theoretical principles must be matched by a corresponding development of skill in the laboratory. Unfortunately students do not always appreciate fully the importance of good laboratory work, carried out intelligently and with a proper understanding of the objectives and principles involved. The difficulty is increased by the fact that many students have little or no opportunity for doing organic chemistry in the laboratory until after they have done a considerable amount of practical inorganic chemistry where the initial emphasis is on analytical procedures, in which a modest degree of superficial success can be achieved without much comprehension of the basic principles. The techniques of preparative organic chemistry make greater intellectual demands from the very beginning: no real progress can be made by attempting to carry out even the simplest preparation as a mechanical routine, and for effective work it is essential to have a sound understanding of the objectives of each step and the physico-chemical principles underlying the methods available for achieving the desired results.

This book aims, therefore, to provide first-year university students and others in schools and colleges who have no previous experience of preparative organic chemistry with a detailed guide for carrying out the procedures commonly needed. Advice and instruction are given on how to do the job, but these are always preceded by discussion of the underlying principles. Specific preparations or reactions are not considered—the emphasis is entirely on those operations normally used for *any* preparation.

Although students need to prepare many organic compounds to illustrate a variety of chemical reactions and principles, this involves the repeated application of the same few physical procedures such as distillation and crystallization. These are known as "Unit Operations" and form the subject-matter of this book. They are discussed here under the general headings:

(a) unit operations involved in carrying out a reaction, and
(b) unit operations involved in isolating and purifying the desired product.

Although preparative organic chemistry utilizes only a small number of unit operations, they cannot be applied indiscriminately as a standard drill. Each procedure must be intelligently selected and applied to meet the demands of the particular preparation; this can only be done with an appreciation of the scope and limitations of the method, which must in turn depend upon an understanding of the principles involved.

All this emphasizes the fact that preparative organic chemistry is essentially an intellectual exercise: manual dexterity without thought or intelligence is useless. In the author's experience this is the main obstacle to be overcome by students first starting this work—they simply do not give enough thought to what they are setting out to do, and the best way to do it. Once they have developed the habit of thinking, and of remembering and applying techniques they have learned in other fields such as gravimetric inorganic analysis, they are well on the way to becoming competent.

Although the emphasis in this book is on unit operations, other aspects of good laboratory practice are also discussed; these include hazards, the importance of yields, and the writing of laboratory notebook records and reports.

Because this book is intended for first-year students, the discussion is limited to the techniques most commonly used. In some cases, however, more advanced techniques are mentioned even though they are not discussed in detail.

In preparative organic chemistry there is no absolute criterion of good practice. Opinions may differ about the best way to carry out the procedures reviewed here, and experienced chemists may not always agree with the recommendations made, but the suggestions given will provide the basis for an acceptable standard of professional practice in the laboratory.

My thanks are due to Mr. M. Mobley, Senior Laboratory Technician, Dyson Perrins Laboratory, University of Oxford, for his assistance in preparing the diagrams.

Oxford
December 1966

W. SABEL

CHAPTER 1

GENERAL INTRODUCTION

The Nature of Organic Reactions

Although the line of demarcation between organic and inorganic reactions is not always entirely clear, organic chemistry can nevertheless be treated as the chemistry of the covalent bond. Ionic species are not frequently involved, and when they are, no special manipulative problems arise.

The classification of compounds as covalent or ionic must be treated with some reserve. There is no such thing as a purely covalent or purely ionic bond between two atoms of different elements; all that can be said is that the bonds in a molecule such as methane are predominantly covalent, while the sodium chloride crystal comprises an aggregation, not of sodium chloride molecules, but of sodium ions and chloride ions, although even here the bonding forces between the sodium and chloride entities are no more than predominantly ionic; there is still some covalent character.

There are some features characteristic of all organic preparations. The materials normally encountered have physical properties associated with the covalent bond, and are usually gases, volatile liquids or low melting-point solids soluble in covalent, non-ionic liquids, in contrast to the inorganic compounds, which, because of their ionic character, are usually high melting-point solids which dissolve in polar (ionic) solvents.

The volatility and low melting-point characteristics of covalent compounds are all explicable on the basis that in these substances the individual units are discrete molecules, held together by relatively weak van der Waals forces. In contrast to this, ionic materials contain electrically charged species (ions), which are held together by much stronger electrostatic forces. In all cases the physical form of a substance is a measure of the "randomness" of its constituent molecules or ions. The conversion of solid to liquid, and liquid to gas, requires energy input because these successive changes of state involve an increasing separation of the component units, whether they are molecules or ions, and this necessitates overcoming the inter-molecular or inter-ionic binding energies.

Organic reactions are usually slower than ionic ones. This is because most inorganic reactions merely involve the formation of ion pairs by mutual electrostatic attraction of oppositely charged particles, a process which, because of the mobility of the ions in solution, is virtually instantaneous. Although a variety of different mechanisms are possible, organic reactions can all be regarded as resulting essentially from electron shifts induced by the reaction environment, leading to the breakage of covalent linkages. This introduces certain reaction characteristics. In ionic reactions the necessary energy is "built-in" by virtue of the existing electrostatic charges, but for organic reactions the electron shifts and resulting bond rupture effects needed as a preliminary to the formation of new bonds are slow processes, requiring the input of energy (usually as heat) for a relatively long period of time, which may range from seconds to weeks. Another characteristic follows from this; for an organic reaction to occur it is usually necessary not only to supply energy in the form of heat, but also to provide special environmental conditions, such as a source of protons added, for example, as sulphuric acid. In the main, because of the rather complex electron shifts involved in organic reactions and their associated energy requirements, there is the possibility of several different routes being followed, all requiring somewhat similar environmental conditions. The result of this is that organic reactions can, and often do, give a multiplicity of products. Also, for similar reasons, equilibrium reactions are frequently encountered, so that again it is impossible to obtain a quantitative yield of the desired product.

In a reaction represented by the equation $A + B = AB$, the formation of each molecule of AB must be preceded by the collision of A with B, but, of course, not every collision will result in a reaction. It is obviously essential therefore to provide an environment for the reaction that makes A and B sufficiently mobile to enhance the possibility of collision between them. The conditions prevailing in a solid substance represent a minimum of mobility of the constituent species, and are therefore least conducive to the collisions required before reaction can occur. Thus, reactions do not normally occur easily in the solid state. For an organic reaction, it would appear to be possible to meet the difficulty by applying heat to melt the solid reactants; this is sometimes done, but usually a solvent is used to provide the necessary liquid phase. Gas phase reactions are also quite feasible, but are relatively uncommon in elementary preparations.

The choice of the type and quantity of solvent used in a reaction depends upon many factors, including its chemical compatibility with the other materials present, and ease of separation of the reaction product. In some cases, the solvent may be chosen to provide certain chemical characteristics, such as acidity or basicity.

It is very often necessary to impose temperature limitations on an organic reaction; a suitable choice of solvent can facilitate this and help also to dissipate heat liberated in an exothermic reaction. Thus, if the desired reaction temperature is 80°C, this can easily be achieved by using a solvent such as benzene which boils at that level; the temperature cannot then rise above the boiling point, and any heat liberated in the reaction will be absorbed as latent heat of evaporation of the solvent. In some cases the use of the appropriate solvent in suitable quantity can affect the course of a reaction and possibly avoid the formation of unwanted by-products.

Even under optimum conditions, in the majority of cases the yield of the desired compound is less than 100 per cent of the theoretical quantity; the reaction may not go to completion and/or side reactions may occur, resulting in the loss either of reactants or the required reaction product. Thus, at the end of the reaction, the isolation of the desired product necessitates its separation from what may be a large number of other compounds. Many of the techniques of organic chemistry are related to that problem.

Basic Principles of Preparative Organic Chemistry

It cannot be emphasized too strongly that all preparative organic chemistry involves two main problems:

(1) How is the product to be made?
(2) How is the product to be isolated in a pure condition from its reaction mixture?

In the early stages of organic chemistry students are apt to concentrate on the first of these, but the second is frequently the major problem, demanding the most skill.

The problem of how to deal with a reaction mixture to extract the maximum amount of the desired product in the highest degree of purity requires considerable thought before starting the reaction. This is a particular illustration of a general principle; successful work in practical organic chemistry always requires the ability to think ahead, not only to the next stage but to the operations beyond that as well. Consideration in advance of how a reaction mixture is going to be treated in order to extract the reaction product, can affect decisions about the way in which the preparation is to be carried out, and the materials to be used for it.

It is sometimes convenient to consider the problem of separation in two stages—the isolation of the main product in a reasonable degree of purity, and the final task of purifying this crude material. In the majority of elementary work in practical organic chemistry, separation operations are the most exacting part of the job, involving many physical techniques and some chemical methods. Physical methods are typified by the use of filtration for separating a solid from a liquid, while chemical operations take advantage of the fact that the physical form and properties of an organic compound can be profoundly changed by a simple chemical reaction, which for this purpose must be easily reversible. Thus, benzoic acid is only slightly soluble in water, but dissolves very readily in sodium hydroxide; addition of a mineral acid to the solution of the sodium salt causes precipitation of the benzoic acid.

The foregoing example relating specifically to benzoic acid leads to another highly important concept of practical organic chemistry. The example given would have been equally valid if reference had been made to toluic acid. From the chemical point of view, this is because organic chemistry does not so much involve a study of a large number of individual compounds as of classes of compounds, having similar properties by virtue of their common functional groups. Thus, the principle used for extracting benzoic acid from ether into water by conversion to the sodium salt can be applied to many other compounds containing a —COOH functional group.

Unit Operations

Purely physical separation methods involve the concept of unit operations. Thus, the process of filtration may be effectively applied for the separation of barium sulphate from water, or of naphthalene crystals from alcohol: in all cases, where a solid is in contact with a liquid phase, separation by the unit operation of filtration is possible, regardless of the chemical characteristics of the system. Similarly, a mixture of two liquids, one of which is more volatile than the other, can usually be separated by fractional distillation, which is yet another unit operation.

In this book the problems of practical organic chemistry are discussed from the viewpoint of the principles and applications of some of the common unit operations, which are considered approximately in the order in which they are likely to be carried out in the laboratory. After a general discussion about apparatus and hazards, consideration is given to the problems involved in carrying out preparative reactions from the viewpoint of such typical unit operations as materials handling and transfer, as well as those such as heating, cooling, mixing, etc., which are involved in achieving specific types of reaction environment. Then the discussion turns to the unit operations involved in isolating the desired compound in a reasonable state of purity.

The methods available for determining the purity of organic substances and criteria for assessing the results obtained are also examined.

Reports and Communications

The growth of a science such as organic chemistry is the result of the activities of large numbers of chemists throughout the world, but this in itself is not enough—development of the subject requires effective communication and exchange of information. As this book is intended for students who are relatively new to practical organic chemistry, it is premature to consider the preparation of papers for publication in the journals of learned societies, but communication at all levels must start with every worker having a complete and accurate record of what he has done and the results obtained: it is therefore essential that all work done in the laboratory should be suitably recorded in laboratory notebooks or files. It is never good enough to depend on memory.

Good Laboratory Practice

With the observance of a few well-defined rules of safety and by the application of general common sense, practical organic chemistry is not an especially hazardous occupation, but careless handling of inflammable or toxic materials may lead to accidents having serious consequences. The nature of the hazards likely to arise and recommendations regarding the methods of dealing with them are therefore discussed in Ch. 3.

In order to acquire a reasonable standard of competence in practical organic chemistry and work efficiently and effectively, it is necessary to develop a suitable mental attitude and to become so accustomed to certain habits of working that they become almost instinctive. Some of the more essential requirements are listed below:

(1) Read the instructions carefully and completely before starting the experiment. A student was once found weighing out a

large quantity of sodium and explained that he proposed to add this to an aqueous reaction mixture in accordance with instructions set out in a printed sheet. He had failed to observe that the words "add 40 g of sodium" at the end of one line were followed by "hydroxide" at the beginning of the next.

(2) Examine the quantities of all reactants required; if necessary calculate the molar proportions.

(3) Having read the instructions, relate them to the chemistry of the reaction, and try to visualize what each step involves and how it will behave.

(4) Consider the possible hazards involved; this includes such factors as the inflammability or toxicity of the reactants or products, and the possibility of the reaction becoming so vigorous as to be uncontrollable.

(5) If specific information is not provided, assess the quantity of material to be handled at every stage, and choose apparatus of suitable capacity.

(6) Before starting the experiment, work out an approximate time schedule for all of the operations: if it then appears to be necessary to leave a preparation unfinished at the end of the day's work, review the situation very carefully, to determine at what stage this can best be done, so that the benefit of work already completed is not lost. If possible leave the preparation where it can profitably continue unattended. For example, if a substance is to be recrystallized, and the operation cannot be completed by the end of the day, work up to the point where the hot solution is left to cool and deposit its crystals. Undesirable changes may take place in a reaction mixture left unattended at an unsuitable stage. For example, if an ester has been extracted and is standing in presence of aqueous mineral acid, it would be undesirable to leave it because of the possibility of hydrolysis. Similarly, if an amide has been obtained in an aqueous alkaline medium, the separation of the amide—usually by filtration—must be completed without delay to remove it from contact with the alkali, and so minimize the possibility of hydrolysis on standing overnight.

Arrange the order of working to minimize waste of time. If, for example, a reaction mixture is to be heated under reflux for 1 hour, nothing is gained simply by watching it continuously. Once the mixture is seen to be refluxing steadily, it can be left to look after itself, and the time devoted to other jobs, such as entering results in a notebook, taking melting points, washing apparatus, etc. In some cases, it may even be best to start another experiment.

This planning and looking ahead is very important, especially under examination conditions, when it is usually found that the required amount of work can only be carried out properly in the period allowed if no time is wasted. In all cases, a few minutes spent reviewing the whole job and studying the instructions can eliminate wasting time while water baths are heated to boiling, or hot solutions left to cool.

(7) Clean working conditions are always desirable, and all spilled materials should be mopped up without delay. This applies not only to one's own working area in the laboratory, but also to such communal apparatus as balances and ovens. The instruction "leave things in the condition in which you would expect to find them" is none the less appropriate for being rather trite.

(8) All apparatus and chemicals for which there is a regular location should be returned there immediately after use. A lot of time can be wasted in looking for things that are left lying around in the laboratory instead of being returned to their proper place.

Yields and Losses—Their Calculation and Importance

In any chemical reaction the yield is limited by the stoichiometry; this can be demonstrated by reference to the formation of ethyl acetate in accordance with the following equation:

$$\underset{60}{CH_3COOH} + \underset{46}{C_2H_5OH} = \underset{88}{CH_3COOC_2H_5} + \underset{18}{H_2O}$$

This shows that 60 g of acetic acid cannot give more than 88 g of ethyl acetate, regardless of the amount of ethanol available, and conversely 46 g of ethanol can only give a maximum of 88 g of ethyl acetate, however much acetic acid is used. In practice, the yields obtained in organic preparations are usually less than the maximum theoretically obtainable.

CALCULATION OF YIELDS

Yields are always expressed as a percentage of that theoretically possible.

$$\text{Yield} = \left(\frac{\text{Weight of product obtained}}{\text{Weight theoretically possible}} \times 100\right) \text{per cent.}$$

The yield theoretically obtainable is not always entirely clear. Reverting to the reaction in which ethanol is made to react with acetic acid, no problem would arise if equivalent quantities of all the reactants were used; if the reaction mixture simply contained 60 g of alcohol and 46 g of acetic acid, the equation shows that the theoretical yield of ethyl acetate is 88 g, regardless of whether one calculates by reference to the acetic acid or the ethanol. In practice, however, it is customary to use an excess of one or more of the reagents. In this example a substantial excess of alcohol is present, because it serves as a diluent as well as a reactant, but the maximum possible yield of ethyl acetate is still limited by the acetic acid. Calculation of yields, both theoretical and actual, are always made with reference to the reactant which is not present in excess; this is one of the reasons for recording not only the weights, but also the molar proportions of all the components of the reaction mixture.

The calculation of yield, which must take account of molecular weights of reactants and product, is based on the following:

$$\text{Yield} = \left(\frac{\dfrac{\text{Wt. product}}{\text{Mol. wt. product}}}{\dfrac{\text{Wt. reactant}}{\text{Mol. wt. reactant}}} \times 100\right) \text{per cent}$$

$$= \left(\frac{\text{Wt. product}}{\text{Wt. reactant}} \times \frac{\text{Mol. wt. reactant}}{\text{Mol. wt. product}} \times 100\right) \text{per cent.}$$

This calculation may be done with a slide-rule. It is quite unrealistic to attempt calculations of yields to more than the first decimal place, because of the inherent limitations of weighing and of carrying out the normal operations, where even with the best technique the reproducibility of the various procedures cannot be such as to warrant any more precise calculation.

It is almost invariably found that preparations in organic chemistry give yields significantly less than 100 per cent of the theoretical, and it is essential to consider the reasons for this. If, for example, a yield is 65 per cent of theory, the question must be asked "what happened to the other 35 per cent?" There is no room for complacency; although 65 per cent is quite a good yield for many student exercises, it is salutary to remember that the lost 35 per cent represents more than half of what was obtained.

In evaluating the results of a preparation in terms of the amount of material *not* obtained, the following questions must be considered:

(1) Was the lost material ever produced in the reaction?
(2) Was the missing material produced in the reaction, but subsequently lost in the process of working up?

With regard to (1), the reaction conditions may have resulted in chemical equilibrium such as in a typical esterification, so that a complete conversion of the reactants to the reaction product is not possible. Alternatively, side reactions may have occurred, with formation of unwanted material, or the desired product may itself have undergone further reaction. A poor yield resulting from incomplete conversion of the reactants because of equilibrium conditions, or possibly because insufficient time had been allowed, has the advantage that the reaction mixture will contain some of the reactants, which may be economically recoverable; if an equilibrium reaction had been involved, the desired product would tend to revert to the reacting materials and the working-up procedure must take account of this possibility. Side reactions are more costly, because they represent an inevitable loss of both reactants and reaction desired product.

Loss of material during the actual working-up operations can be considered in stages; it may be due to physical causes such as spilling or solubility, and the simple process of transferring a material, especially a liquid, from one container to another will involve losses. Ways of minimizing drainage losses are described on p. 44. Another lesser source of loss is chemical change, such as polymerization of the product while it is being isolated. Thus an ester may be hydrolysed if left in contact with alkali. Failure to make the desired product in a reaction is a question mainly of chemistry, and therefore falls outside the scope of this book, but problems of isolating the material once it has been made are a matter of technique.

THE ARITHMETIC OF YIELDS AND LOSSES

The importance of overall yield or losses—they are always related—must be seen in its proper perspective. First consider the way in which losses accumulate. Even a simple, one-stage chemical reaction, such as the esterification of benzoic acid with propanol, involves several manipulative stages each involving some loss, and the position becomes correspondingly serious when multi-stage preparations are done. Consider, for example, the conversion of aniline to *p*-nitroaniline by the following reactions:

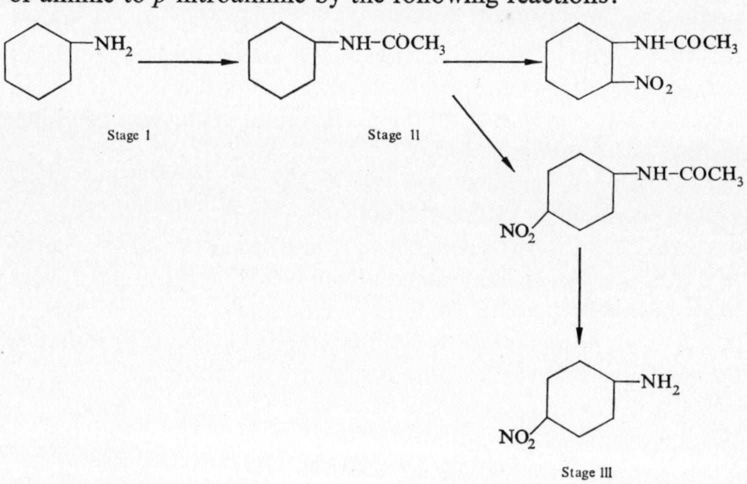

For any one chemical stage an overall yield of, say, 90 per cent of theory is usually very good; in the above sequence, the overall yield from these reactions would be 90% of 90% of 90% = 73%. That is to say, 27 per cent of the yield theoretically obtainable has been lost.

In many reactions, series yields for each stage may be much less than 90 per cent. If each stage gives only a 50 per cent yield, the above three-stage reaction would give an overall yield of 12·5 per cent of theory. Five stages at 50 per cent for each give an overall yield of 3 per cent; that is, 97 per cent of the potential yield has been lost.

The importance of this simple arithmetic cannot be overstressed, but there are two aspects which are especially worthy of comment. A considerable amount of research work is done on certain natural products obtained by extraction from plant or animal tissues: suppose that the tissue in question gives 5 per cent by weight of an extract which can be regarded as the starting material for carrying out five subsequent chemical stages, each giving a yield of 50 per cent of theory. The overall yield, having obtained the starting extract, is only 3 per cent of theory, and the weight of product finally obtained from 1 kg of tissue may be only about 0·5 g (the exact figures depend upon molecular weights). This does not leave much for further stages, and handling 1 kg of plant or animal tissue for extraction is itself a bulky operation. In natural products especially, therefore, yields can be so low that a stage is rapidly reached when milligram quantities are involved in a reaction. It is therefore all the more important to apply suitable techniques to minimize losses.

Finally, in this context, it is interesting to look at the opposite side of the picture, and consider the skill required to produce an overall yield of, say, 10 per cent in a multi-stage synthesis, involving, for example, eighteen stages. Such yields on this type of work have been obtained in some industrial operations, and simple calculation shows that the individual stage yields must be well over 90 per cent, leaving very little room for any losses.

CHAPTER 2

REVIEW OF APPARATUS AND EQUIPMENT

Glassware

Most laboratory glassware is now made of borosilicate glass, such as Pyrex or Hysil, although a certain amount of soda glass apparatus is still used.

Borosilicate glass has several advantages over soda glass. It is less prone to chemical attack, and the coefficient of expansion is lower, so that it is more resistant to thermal shock. When a piece of glass is heated on one side only, local expansion takes place, and because the rate of heat transfer through the glass is low, a steep temperature gradient is set up, which produces a corresponding tensile stress. This can cause breakage, because although glass is strong under compression, it is much weaker under tension: borosilicate glass can therefore better tolerate uneven heating, so that thicker and hence mechanically stronger vessels can withstand the same degree of thermal shock as thinner soda glass. Alternatively apparatus made of borosilicate glass can withstand a wider range of thermal strain than soda glass equipment of the same thickness.

Although soda glass lacks resistance to thermal shock, it is sometimes more convenient to use it in the laboratory for such purposes as making melting-point tubes, etc., where its relatively low softening point makes it easy to draw out in an ordinary bunsen burner flame, and facilitates other glass-working operations. Glass-blowing with soda glass, however, is not easy in spite of these advantages because, having heated the glass, it must be very carefully annealed to eliminate strain during cooling. Failure to do this will inevitably result in fracture. After making any kind of fused joint in soda glass, it must only be allowed to cool very slowly and evenly, and to facilitate this the hot glass surface should be coated with a film of soot by deposition from a luminous bunsen burner flame or a wax taper.

Borosilicate glass can only be worked satisfactorily with the use of an oxygen–coal gas flame provided by a blowlamp or torch; an ordinary air–coal gas flame, such as provided by a bunsen burner, is not enough for more than simple bending.

It is not possible to make satisfactory fused joints between borosilicate and soda glasses because of the differences in thermal expansion properties. Such joints will fail during cooling. It is sometimes possible to distinguish the two types of glass by visual inspection; the cut end of a piece of soda glass has a white or greyish-green appearance, while borosilicate has a distinct yellowish tint. If a sample is heated in a bunsen flame, soda glass will quickly become white hot and soften, whereas borosilicate glass will merely become red hot, and unless the sample is very thin, it will not easily bend or collapse.

Whenever glass tubing or rod are used, it is very important to flame-polish the ends in order to remove the sharp edges. This is done by heating the end of the tube or rod in a roaring bunsen flame until the sharp edges of the glass become red hot. The glass should be rotated between the finger and thumb while being heated. Excessive heating may cause collapse or distortion of the glass. Soda glass rod should not be placed directly into a roaring bunsen flame, because the high coefficient of expansion will probably result in shattering; it should be cautiously introduced into a non-roaring flame, and the air supply gradually increased to raise the temperature.

It is often necessary to bend glass tubing and rod. This operation is quite easy, and the most important requirement is patience. The tube is rotated continuously in the bunsen flame, so that the maximum length, at least 5 cm, is heated until it is red hot: the tube is then removed from the flame, and bent to the desired angle with one quick, steady movement. This technique applies to any size of tubing up to about 5 mm outside diameter, which is enough for most purposes. If the heating is applied gradually and the glass is not allowed to become too soft, the bends obtained should be smooth and neat, with the bore virtually unchanged. It is difficult to obtain a satisfactory bend of more than 90° with larger sizes of tube. If a U-bend is required, the best procedure is to make two right-angle bends as close together as possible, but in separate operations.

The same principles apply to soda as to borosilicate glass, but the glass must not be allowed to become too soft; if it does, the tubing will almost certainly collapse on bending. While borosilicate glass requires a roaring bunsen flame fitted with a bat's-wing spreader, soda glass may be softened excessively under similar conditions. Soda glass should be just about red hot, and the bend made before the glass shows any serious tendency to droop.

In the discussion about techniques of vacuum distillation, reference is made to the use of a very fine glass capillary to promote smooth boiling (see p. 107). It is best to make this from a length of thick-walled glass tubing (internal diameter less than 1 mm, outside diameter 4–5 mm) which is rotated and heated strongly so that the minimum length—about 2–3 cm—softens. When the glass has become very soft, it is removed from the flame and the ends slowly pulled apart about 10 cm to give a thick-walled capillary between the unheated portions; it is then held by one end so that it hangs down vertically. In this way, the capillary does not sag, and remains in the same axis as the undrawn portion. The process is then repeated, heating the centre portion of the capillary, and drawing out to give a very fine thread of glass. It is rarely possible to draw the glass out too finely: the usual fault is to leave the capillary much too coarse. When drawing out borosilicate glass tube, an oxygen–coal gas blow-pipe with a fine pencil-type flame should be used, but a roaring bunsen flame is sufficient for soda glass. For preparing capillary tubes for the determination of melting points, ordinary narrow-bore glass tubing is treated in a similar way.

Ground-glass Fittings

Ground-glass fittings in the form of stop-cocks fitted to burettes and separating funnels, etc., require lubrication, to obtain smooth sliding movement of one ground surface over another. The lubricants must be selected so that they are not dissolved or otherwise attacked by the liquid with which they come into contact when the stop-cock is in use. Petroleum jelly is satisfactory for use with aqueous solutions, but quickly fails in contact with organic solvents, such as benzene: when this failure occurs, not only does the stop-cock tend to bind or leak, but also the lubricant dissolves and hence contaminates the solvent. Special greases, based on lithium stearate, or silicones, are used in contact with organic solvents. For very severe conditions, such as exposure to undiluted bromine, syrupy phosphoric acid is an effective lubricant.

The state of lubrication of a stop-cock should always be examined before the apparatus is put into use. If the plug does not turn easily and smoothly in the bore, lubrication is obviously necessary; if there is an incomplete film of grease on all of the sliding surfaces, they have a characteristic streaky appearance, and the streaks appear to move when the plug is rotated in its socket. If there is any doubt at all, the lubricant must be renewed before putting the stop-cock into use; this simple precaution can obviate considerable subsequent delay and irritation. A wet or dirty surface cannot be lubricated satisfactorily; the plug must be removed from the bore, and both ground surfaces thoroughly cleaned and dried by wiping with a duster or a paper tissue. Two thin films of grease are then applied to the plug only, so as to form a pair of narrow parallel rings, one on each side of the hole. The plug is then refitted into the clean dry bore, pressed home and turned in order to spread the lubricant into a thin, even film. Excess of grease should be avoided, because of the risk of blocking the hole with the surplus.

When using a stop-cock, it is always necessary to rotate the plug with a slight pressure applied inwards to offset any tendency to lift it out of the tapered bore. This is especially important when the stop-cock is carrying liquid, because if the plug should lift even momentarily and permit liquid to enter between the lubricated ground glass surfaces, it will be found impossible to continue using the stop-cock without re-lubricating it.

If a stop-cock does not have a retaining device to prevent the plug becoming detached from the socket, a small rubber band will provide this safeguard. This must be fitted and used with care, because repeated turning of the plug in the same direction may cause the rubber to wind up and act as a torsion spring, so that the stop-cock plug rotates when left unattended, especially if it has been well lubricated with a low-viscosity grease. The best safeguard is to remember to turn the stop-cock plug alternately in a clockwise and anti-clockwise direction, to prevent the possibility of the rubber band winding up.

STANDARD TAPER GROUND-GLASS JOINTS

Borosilicate glass is almost always used in the construction of ground glass joints which are made by fitting a cone (male) into a socket (female). Most manufacturers have standardized the dimensions of these joints, so that they are completely interchangeable: the taper is 1:10, and they are available in a series of standard diameters. For most elementary laboratory work the joint sizes most commonly encountered are B19 and B24, but smaller joints, such as B10 and B14 are also used. The prefix "B" indicates the length of the ground glass portion of the joint: the "A" series, which are longer than the "B", are occasionally used in high vacuum work. Adaptors are available to permit the joining of cones and sockets of different sizes—for example, a B19 cone into a B24 socket.

Lubrication of well-made standard taper joints is not usually needed if they are used under atmospheric pressure and are not required to rotate during the experiment. For vacuum work, especially high vacuum, a suitable low vapour-pressure grease should be used. Ground joints used in the presence of strongly alkaline solutions, even under atmospheric conditions, may sometimes tend to seize, and to avoid this, light lubrication with a hydrocarbon grease may sometimes be desirable. It is good practice to dismantle ground glass joints as soon as possible after use, especially if they are exposed to strong alkali.

In extreme conditions, where no ordinary lubricant is suitable, seamless conical sleeves made of thin films of polytetrafluoroethylene (P.T.F.E.) are available, which fit accurately on to the cone, and provide a non-sticking interface between the ground glass surfaces of the mating joints.

In some complex assemblies, problems of alignment can be overcome by the use of ground-glass joints of a hemispherical shape. These give a ball and socket type of joint, held together by a spring clip device. They can take up a wide range of angular movement without impairing the sealing properties of the ground surfaces, and so can accommodate a considerable amount of misalignment in an all-glass assembly.

Corks and Bungs

Corks and bungs are available in a range of sizes, usually with tapering sides, and the dimensions are graded, so that the large diameter of one cork or bung has the same dimensions as the smaller end of the next larger size. Corks of the range of size commonly used in the laboratory—up to about 4 cm in diameter—are cut from the bark in such a way that the fissures in this naturally-occurring material are parallel to the circular end faces. Thus the maximum diameter of the cork is limited by the thickness of the bark.

Corks are rather hard when new, and do not easily fit to give a gas- or liquid-tight seal in the neck of a flask. The best way of softening a new bark cork is to place it on its side on the floor and roll it under the sole of the foot, using light pressure and keeping the cork clean by placing it between two sheets of paper; this is a very simple technique that usually gives much better results than the machines commonly bought for the purpose.

BORING CORKS AND BUNGS

It is often necessary to bore holes in corks and bungs. This operation is quite simple and satisfactory provided that the borer is really sharp, with a clean, even cutting edge, and if the cutting operation is about 90 per cent twist and 10 per cent push. The borer must be well lubricated by dipping into glycerol or methylated spirit. When boring the hole, if the cutter appears to "drag" it should be removed, and re-lubricated before resuming the operation. Most failures are caused by excessive pressure and/or a blunt cutter; the cutter should never be used as if it were a punch. In all cases the aim is to bore a hole of the right diameter in the right position and with walls that are smooth and even. This is especially important if a rubber bung is bored as part of a high vacuum assembly; failure to do the job properly will result in air leaks when the system is under reduced pressure. If excessive pressure is applied to the cutter, the rubber will be distorted and the hole will not have smooth sides or an even bore. The choice of the borer size needs care; for a cork a borer should be chosen a size smaller than one that will just slip over the glass tube or rod to be fitted in the hole to be bored. If the resulting hole in a cork is slightly too small, it may be enlarged by the use of a small rat-tail file. The cork should always be softened *before* boring. When boring a rubber bung it is sometimes advantageous, especially for high vacuum work, to use a borer one size smaller again, to make a vacuum-tight seal. This is possible because of the greater flexibility of the rubber.

The cork borer must be very sharp and have a smooth cutting edge; before attempting to sharpen a cutter having a rough, jagged edge this must first be smoothed by grinding on a piece of carborundum paper laid on a hard, flat surface. The grinding must be continued until the edge is quite smooth, and the end face perpendicular to the main axis of the borer. The flattened, blunted end is then easily sharpened with the usual type of hand tool supplied for the purpose, lubricating the cutting edge by dipping the end of the cork borer into glycerol. This lubrication should be renewed several times during the sharpening procedure. Undue pressure on the cutting blade of the sharpener must be avoided, and only very gradual removal of surplus metal should be attempted.

Even if the main part of boring the cork or bung is done properly, the effect is frequently spoiled at the last stage. It is not possible to obtain a satisfactory result by trying to bore from both ends of the cork or bung. Just before the cutter emerges the bung or cork should be pushed hard on to a wooden or cork support, and the cutting process completed even more carefully than before, with less pressure.

It is essential to provide lubrication whenever glass rod or tubing is inserted into rubber tubing or a cork or bung. Mineral oil is not suitable, because it may attack the rubber; one of the best lubricants is glycerol, but ethanol or even water may be used. It is an advantage if the lubricant can be removed after use simply by washing with water. Jagged ends of glass rod or tubing must be flame-polished before insertion into a rubber or cork bung: if the edges are sharp, the bore will almost certainly be cut, and a leaky connection will be the result. To flame-polish a cut-end of rod or tube, the glass is rotated in a roaring bunsen flame until it becomes red hot. This is sufficient to cause the sharp edges to become rounded without allowing more extensive distortion of the glass.

Size of Apparatus

FLASKS

As a general rule, a flask in which a reaction is being carried out should be not more than two-thirds full. This leaves sufficient free space above the liquid surface to accommodate normal boiling or stirring effects, without risk of liquid rising into the neck of the flask. Some reactions require an unusually large free space; if, for example, the reaction is known to be very vigorous and accompanied by severe frothing, e.g. a Grignard reaction, it is obviously desirable to have plenty of room above the liquid, and the flask capacity should be chosen so that the reactants initially occupy no more than about one-third of the capacity. If a reaction does not involve boiling, or where operating conditions are relatively quiescent, as in a precipitation, it may be possible to use a flask three-quarters full.

For a distillation at atmospheric pressure the flask should be about two-thirds full at the start of the operation, but for vacuum distillation the boiler flask should not be more than half full initially.

SEPARATING FUNNELS

When separating funnels are required for extractions, as distinct from their use as a means of controlled addition of liquid from one container to another, room must be left for efficient mixing of the two liquid layers when they are shaken. The funnel should not be more than three-quarters full when both liquid phases have been added.

Cleaning and Drying Glassware

CLEANING

Cleaning apparatus can be regarded as an exercise in solubilities. Although some of the contaminants can be removed by scraping with a spatula or by some other mechanical means, the elimination of last traces from the glass surface is usually only possible with a solvent or dispersant. Water containing a soap or synthetic detergent is a good solvent for an enormous range of materials, and even where it is not able to dissolve the "dirt", it can act as a suspension medium which, aided by the detergent, gives a colloidal system. The cleaning of glassware contaminated by organic materials presents problems, because it is necessary to remove those substances which, on account of their covalent character, are least soluble in water. Thus, water alone is not often suitable as a cleaning agent, although some detergents of non-ionic type are available that considerably extend its effectiveness.

The addition of aqueous acid or alkali will increase the solubility of an organic substance in water where salt formation is possible. Hydrochloric acid is best for removing an organic base; sulphuric acid may form much less soluble salts; nitric acid must never be used for dealing with organic materials, because it may cause violent and dangerous explosions. Sodium hydroxide or sodium carbonate are widely used to produce alkaline cleaning conditions, and if they are not satisfactory, it is unlikely that other inorganic bases will be any better.

Ethyl alcohol is an excellent solvent for a wide range of materials, and its miscibility with water makes it especially useful. There is no justification for using high-grade alcohol for cleaning purposes: it can be a useful economy to use recovered or contaminated material if this is available. If several flasks require cleaning with a solvent, waste can be avoided by using one portion and transferring it from one flask to another. Ethanol in the form of industrial methylated spirit (IMS) is most commonly used.

DRYING GLASSWARE

It is often necessary to use apparatus that has been properly dried. Sometimes reaction conditions, such as the Grignard reaction, require this and in other cases the presence of water may cause undesirable solubility effects.

In most laboratories, provision is made for drying washed apparatus in a suitable oven; if this is used, it must be remembered that water is relatively slow to evaporate because of its high latent heat, and sufficient time must therefore be allowed. Several hours may be needed, and it is not good enough simply to leave the wet glassware in the oven for a few minutes. Electrically heated ovens with exposed elements may be provided, and then the possibility of fire and explosion arises if glassware wet with inflammable solvents such as ether is introduced.

Glassware wet with water may be dried quickly and effectively by the use of some solvents. After draining off as much water as possible, the apparatus may be rinsed twice with a small volume of acetone (e.g. 10 ml for a 150 ml flask), after which it may be dried by blowing air through it from a pump or rubber blowball. This must be done well away from flames to avoid the risk of fire from the air–acetone mixture blown out. Another drying procedure depends on first rinsing twice with alcohol (methyl or ethyl) followed by two rinses with ether, which is finally removed by air blowing.

If several pieces of apparatus are to be dried at the same time, it is a useful economy to rinse from one container to another with the same portion of solvent. Sintered glassware such as in a Buchner funnel may need to be dried, even if it is not obviously wet with water, and it is usually best to filter the appropriate solvents through it; if this is done under vacuum, not only is the procedure facilitated, but it provides an opportunity of using the same solvents to dry the filter flask as well.

CHAPTER 3

HAZARDS

PRACTICAL organic chemistry involves potential hazards for both persons and property, but these can usually be avoided unless work is done in a slovenly way, or without proper thought. The avoidance of injury to people is the primary responsibility of everyone working in a laboratory, and this requires a clear understanding of the principles and reasons underlying all the procedures used: this eventually becomes almost instinctive. No single textbook can constantly draw attention to all of the hazards involved in a chemical operation, and further reading on the subject is recommended. People are always more important than property. If, in spite of all precautions, an accident should occur, the first essential is to minimize danger or harm to people; the problem of safeguarding buildings and equipment must always take second place.

Toxicity

Ingestion or inhalation are the most obvious ways in which toxic substances may be introduced into the body, but the possibility of some materials entering the bloodstream by passage through the skin must also be considered; this can occur by a process of absorption, even if there is no obvious cut or abrasion. The skin itself can be damaged by contact with a wide range of chemicals, giving rise to the condition generally known as dermatitis.

The ability of some substances to pass through the skin into the bloodstream seems to depend upon their properties as fat solvents.

All organic liquids could be expected to have this property to some extent, but some compounds are especially harmful, e.g. aniline and nitrobenzene; these are absorbed into the blood by passage through intact skin, and can cause serious illness. Any accidental contamination of the skin with these substances must be immediately attended to. Washing with a copious supply of soap and water is the best first-aid treatment. All aromatic compounds, including the hydrocarbons, but more particularly benzene, amines and nitro-substituted materials, should be regarded as toxic by skin absorption, and handled with suitable care.

Dermatitis is a much more complex problem. Phenols and glacial acetic acid cause severe local skin damage, and so can be regarded as dermatitic in the rather loose sense of the term, but many other chemicals may cause skin irritation and inflammation in a rather unpredictable way, depending on individual susceptibilities. Virtually any substance can cause irritation in people with sensitive skins; care should always be taken to minimize contact of chemicals with the skin, and when it does occur, washing with plenty of soap and water should be carried out without delay. The use of solvents other than water to remove materials contaminating the skin is not to be recommended, since the solvent may merely facilitate passage of the contaminant deeper into the skin.

Toxicity may be acute or chronic. Acute toxicity, which is the more common hazard in student laboratories, develops suddenly, and is illustrated in its most extreme form by the very rapid and disastrous onset of illness caused by the ingestion of cyanide. Chronic toxicity is the condition brought about gradually after prolonged exposure to the toxic agent. An example of chronic toxicity is provided by benzene, which, on continued inhalation or exposure, can produce severe damage to the blood-forming system.

It could be argued that virtually all chemical materials, especially organic ones, are toxic in that they can be expected to cause some degree of damage if ingested, inhaled or kept in contact with the body in sufficient quantity, over a sufficiently long period of time. On a more practical appraisal, the following commonly used materials may be regarded as toxic in the normal sense of the term:

CYANIDES

Potassium or sodium cyanides are used in various reactions such as the benzoin condensation. The dangers of handling these materials cannot be over-emphasized and although first-aid measures are normally provided in the laboratory, it must be understood that cyanide ingestion is almost invariably lethal. The only reasonable way of handling them is to assume that if even a very minute quantity is ingested it will cause very rapid death. It should also be appreciated that exposure to atmospheric moisture is sufficient to hydrolyse the simple alkali metal cyanides, which are most commonly used, liberating hydrocyanic acid, which is lethal by inhalation. These cyanides must be handled with the utmost care in a fume cupboard and rubber gloves must be worn. In any case, after handling cyanides, or any reaction mixture containing them, the hands, and likewise all apparatus that may have been contaminated, must be washed with scrupulous care.

ALCOHOLS

Normally these are not particularly dangerous, but methanol can cause severe, irreversible damage to the optic nerve resulting in incurable blindness, if ingested or inhaled for long periods. In normal use in the student laboratory, however, the toxicity of methanol may be regarded as mildly chronic, and with ordinary care it is unlikely to cause trouble.

HYDROCARBONS

Aromatic hydrocarbons are much more toxic than the aliphatics, and benzene is by far the most dangerous. It is interesting to consider the reason for this: in general, the body's defence mechanism for dealing with a toxic substance is to convert it to some other form, less harmful, which can be more effectively excreted. The detoxication mechanisms of the body for dealing with hydrocarbons are to oxidize them in the liver to carboxylic acid derivatives, which are soluble and can be excreted in the urine. Toluene, for example, is first oxidized to benzoic acid; but benzene, lacking a readily oxidizable hydrocarbon side chain, cannot easily be detoxicated in this way, and accumulates in the liver, causing severe damage to the blood-forming system.

CHLORINATED HYDROCARBONS

All halogenated materials, such as chloroform or trichloroethylene, must be assumed to be potentially harmful. They are apt to cause liver damage, and their fat-solvent properties make them able to pass easily into the bloodstream by skin absorption. Inhalation, or prolonged contact of these materials on the skin, must be avoided.

PHENOLS

Many of these are toxic—some are used as antiseptics—but the lower members, especially phenol itself, can also cause blisters if left in contact with the skin. Phenol must therefore be handled with great care to avoid skin contamination; if, however, this occurs, the affected area should be washed very thoroughly with water. It is sometimes recommended that the skin should be washed with dilute bromine water, which neutralizes the phenol by forming the insoluble tribromo compound. Any skin contamination with phenol should receive expert first-aid attention. It is most important *not* to wash with alcohol or other organic solvent, which merely aggravates the trouble by facilitating transfer of the phenol through the skin into the bloodstream.

AMINES

Most amines are toxic. The primary aromatic ones are the worst and aniline, for example, must be treated as a seriously toxic material. It can cause severe damage, even by skin absorption. The special danger of β-naphthylamine, which is carcinogenic, must also be noted, but it is not likely to be encountered in normal student practice.

NITRO COMPOUNDS

These should always be regarded as potentially hazardous. Most of them can produce chronic toxicity effects including progressive damage to the blood-forming system. They can be absorbed through the skin. Apart from toxic effects, some nitro compounds are explosive.

MINERAL ACIDS

Dilute acids are not especially toxic or corrosive, but the concentrated materials must be handled with great care. Nitric acid and sulphuric acid are both powerful oxidizing agents as well as being strongly acidic, and can cause very severe burns. Any spillage, especially on to the skin, must be immediately mopped up and washed with a copious supply of water. Nitric acid can cause serious fires and explosions if mixed with some organic materials. Concentrated sulphuric acid is not quite so troublesome as nitric acid from the viewpoint of its oxidizing properties, but, nevertheless, considerable care must be taken whenever it is introduced into any other organic reaction system. Dilution of concentrated sulphuric acid with water or ethanol is a strongly exothermic process that can cause the acid to splutter. In all cases the acid must be added slowly to the diluent, and kept well mixed. The diluent must never be added to the concentrated acid.

BROMINE

This is the halogen most likely to be encountered and to cause trouble. The liquid can cause severe skin burns: any affected skin area must be washed copiously with water. Subsequent washing with glycerol is also recommended. The vapour can also be very damaging, especially if it is inhaled, or comes into contact with the eyes. Bromine must therefore always be handled very carefully, in a fume cupboard, and any mishap dealt with at once with expert first-aid.

CAUSTIC ALKALIS

Sodium hydroxide and potassium hydroxide are commonly used in organic practical work, and their corrosive effect on the skin and eyes must always be borne in mind.

Bodily Injuries

Because most laboratory apparatus is made of glass, cuts are the main injuries to be expected, although the incidence and seriousness of these can be reduced to minor proportions if due care is taken.

Eyes are most vulnerable, and neglect of simple precautions introduces risks of very severe damage, and even of total blindness. This risk cannot be over-emphasized, and it is quite inexcusable to omit the few elementary precautions that could make the difference between safety and disaster.

Flying glass or chemicals are obviously a source of danger to the eyes, and suitable goggles or some other protection should always be worn when working in the laboratory; this precaution is even more important when working with apparatus at reduced or high pressure. Cutting glass tubing can itself be hazardous. Having cut a file scratch on the glass, the bending action required to break the rod or tube should be done away from the body, in such a way that, if a splinter detaches itself from the broken surface, it will fly away from the face and thus minimize the risk of eye damage. As a further precaution it is a good practice to develop the habit of momentarily closing the eyes just before the actual break of the glass occurs. Similar care must always be taken to protect the eyes from damage by splashes of liquid or particles of solids.

In all cases, even what seem to be the most trivial of eye injuries should be treated as potentially dangerous. If, in spite of precautions, an accident should occur to the eyes, immediate first-aid, and possibly medical attention, should be sought.

Although eye injuries are of special importance, quite severe cuts generally can be caused by the bad handling of glassware. One of the most common risks is in the preparation and manipulation of glass rod and tubing, especially when fitting it into rubber tubes or bungs. Glass rod or tubing should never be used unless the ends are smooth and free from sharp edges. Jagged ends, caused by breakage, are especially dangerous, and should be eliminated by cutting off to give a clean edge, but this will still be sharp and likely to cause cuts, and must be flame-polished as described on p. 14. The process of fitting glass into a rubber bung or rubber tubing can cause very severe cuts, and special care must be taken. This operation must be done by a pushing, twisting motion, and if precautions are not taken, breakage can cause deep, and often very serious, damage to the hands. Accidents of this kind have been known to cause damage to tendons resulting in permanent injury. Two precautions must be taken. The passage of glass through the rubber bung must be eased by suitable lubrication, e.g. by glycerol, as described on p. 20. For passing glass rod through a bark cork, water alone may be used as a lubricant, but methylated spirit or glycerol are better. Without lubrication the frictional resistance is so great, especially when passing glass through a rubber bung or tube, that breakage is almost certain to occur; but even when a lubricant is used, there is always the possibility of accidents, and to prevent injury to the hands, they should be protected with a cloth or duster.

Smoking

Even if smoking in the laboratory is not positively forbidden, it is undesirable, and can be the cause of fire and toxic hazards. Apart from the obvious risk that a lighted cigarette or pipe can cause ignition of some vapours (carbon disulphide is especially hazardous), fires have been known to occur as a result of a cigarette-end left smouldering on a bench or in a refuse bin.

Smoking can also introduce a distinct toxicity hazard, apart from its tendency to cause lung-cancer. Some chlorinated hydrocarbon vapour can be decomposed by coming into contact with the tip of a lighted cigarette to give highly toxic products (phosgene?) which can then be ingested by inhalation through the cigarette. Even if smoking is permitted, therefore, it is essential not to do so if there is any general fire risk, or if handling (or in an atmosphere containing) any halogenated hydrocarbon.

It has been suggested that a lighted cigarette in contact with polytetrafluoroethylene (Teflon, Fluon) can cause liberation of volatile, very dangerously toxic, degradation products.

Fire

This is the main hazard apart from toxicity, but with reasonable care and intelligence, risks are not great. Inflammability is a function of volatility and of the ease with which oxidation can occur at high temperatures. Thus, the fire risk in practical organic chemistry is simply a consequence of the use of covalent (and hence volatile) compounds of carbon, that are easily oxidizable to carbon monoxide, carbon dioxide, water, etc. It is possible to obtain a quantitive measure of inflammability of a substance by reference to its flash point, which is the temperature of the liquid at which it is just possible to cause ignition of the vapour by application of a flame: the lower it is, the more highly inflammable is the substance. In the U.K., all substances whose flash points are below 73°F must by law be labelled "Highly Inflammable", and conditions are laid down regarding the handling and storage of such materials.

In student laboratories, nearly all fires are caused by neglect of simple precautions. Most are the result of liquids either spilling or boiling over and coming near a flame. There are other less likely causes, such as sparking from electrical equipment. The increasing use of electric heating mantles and hot-plates to replace Bunsen burners has tended to reduce the number of fires caused in the laboratory, both by elimination of the bare flame as a source of ignition, and also by preventing the breakage of glass flasks caused by careless heating in a flame.

Bare flames should only be used in an organic laboratory when they are absolutely necessary, and extinguished whenever they are not required. Inflammable solvents should not be handled or stirred near to lighted burners. If an inflammable liquid is being heated, other inflammable materials should be kept as far as possible from it, so that if a fire occurs there is a minimum risk of it spreading. Precautions are especially important if a reaction mixture is to be heated for long periods without supervision; it is essential to start the reaction in good time to permit proper adjustment of heating rate, and to see that the process has settled down before it is left unattended. As reaction proceeds, the condition or nature of the mixture may change, e.g. slow deposition of solids, such as potassium bromide, may occur during the reaction. If no stirring is provided, the solids may settle to the bottom and add to the risk of breakage of the flask by local over-heating, due either to thermal strain, or by bumping or other uncontrolled boiling conditions. This kind of reaction mixture should therefore always be provided with a mechanical stirrer. The fine adjustment of a gas flame required to give precise control of a boiling rate can often be best achieved by using a screw clip fitted to the rubber tubing supplying gas to the burner. In most cases this permits better control than is possible with the usual type of gas tap.

When heating a reaction mixture under reflux, the maximum temperature attainable is the boiling point, and nothing is gained by having a heating rate more than just sufficient to give slow, steady boiling. Under these conditions, the load on the condenser is minimal, and there is reserve capacity to cope with a greater rate of boiling that would result if the gas pressure should increase during the night. This minimizes the risk of the condenser being overloaded, with consequent loss of reaction mixture which might boil out and ignite.

A common cause of fire is failure of a reflux condenser because the water supply is interrupted. Water taps sometimes tend to turn themselves off, probably because of swelling of the washer, and to provide for this possibility the supply through the condenser should be slightly increased when overnight running is required. Excessive water flow, however, increases the risk of rubber tubing connections being blown off, or bursting. Thick-walled pressure tubing should be used for the water connections to the condenser, especially the inlet to it from the tap, and all points where the rubber is joined to glass or to the tap itself should be tied with wire.

FIRE EXTINGUISHERS

The action to be taken in the case of a fire should be the subject of rules laid down by those in charge of the laboratory, and it is essential that all students should make themselves fully familiar with their duties and responsibilities in these circumstances.

Every student should know exactly what types of extinguisher are available and where they are kept. In addition, it is important to develop a habit of thinking in advance about the action to be taken if any laboratory operation should lead to trouble. For example, in a Claisen condensation reaction the ester is not very highly inflammable, but the use of sodium requires special care. Hydrogen will be evolved in the reaction, and so there is a risk of explosion. If a fire should start, it is essential to know the right kind of extinguisher to use. Because of the presence of sodium, the use of water as a fire-extinguishing medium could be disastrous.

If, in spite of all precautions, a fire should start, it cannot be too strongly emphasized that the only immediate and instantaneous demand on everyone concerned is to do nothing at all except move away from the area involved, and think. It can be reasonably assumed that a fire starting in the course of student preparation is unlikely to be immediately catastrophic, and much more harm and damage are likely to be caused by panic-stricken rushing around, looking for and using an extinguisher, than pausing for a few seconds to assess the situation and to decide calmly on the right course of action. The basis of all thinking about hazards in the laboratory and methods of dealing with them is that people are always more important than property. It is absolutely essential to remove any casualties and see that they are properly safeguarded from further harm before worrying about the fire or other mishap.

Several different types of fire extinguisher are available, and the wrong choice in dealing with any specific type of fire may do more harm than good. Although, for most general purposes, water is regarded as the most suitable fire-extinguishing agent, this is not the case when dealing with fires in the organic chemical laboratory. The special risks of using water when the alkali metals are involved have already been mentioned, but there is the further risk that burning water-immiscible materials can be spread by the use of water as an extinguisher.

In general, fire extinguishers act in two ways; they cool the burning material and prevent access of air. The following types of extinguisher are commonly available in chemical laboratories:

Carbon dioxide: for virtually all types of fire.

Carbon tetrachloride or other alkyl halides: for most purposes, except burning alkali metals. Produce toxic vapours.

Soda–acid: useful but messy—not recommended if electrical equipment is involved in the fire. Not for use in extinguishing burning alkali metals.

Foam: useful but *very* messy—limitations as for soda-acid.

Sand: Cheap and effective, but rather messy—mainly useful for small fires. Best for alkali metals.

For most of the small fires normally encountered by students in organic chemistry the first line of action should be to try to smother the burning material with a cloth or other cover. If this is not possible, the most convenient extinguisher, in most cases, is carbon dioxide.

After an extinguisher has been used the fact must be reported to the supervisor. Although some carbon dioxide extinguishers are fitted with valves that are claimed to permit discharge of only a portion of their contents if required, the closures do not always give a leak-proof seal after such use. Even if only a portion of the contents of an extinguisher has been discharged, the appliance must be returned for expert service, and replaced by another one known to be in good condition.

Fire-fighting is a highly-skilled and specialized function, and normally the student would only be directly concerned to the extent of dealing with small outbreaks, but in the unlikely event of a fire getting out of control, outside help must be obtained in accordance with fire regulations laid down for the laboratory.

The student's principal responsibilities are to take all possible precautions to prevent an outbreak, to be fully familiar with the laboratory regulations regarding fires, and, if there is an incident, to give first attention to safeguarding people rather than property.

Explosion

In all explosions the common feature is the sudden liberation of energy. Within the context of this book the type of explosion most likely to be encountered is that caused by a rapid increase in pressure inside a closed system, resulting in failure of the walls of the container. In some cases, no container is involved; this arises when a local concentration of hydrogen or organic vapour in the air is ignited by a spark or flame. This is essentially an oxidation reaction, and the energy liberated can cause severe damage. In student work this explosion is more likely to take place in a glass container which will probably be shattered; this obviously makes the consequences much more serious. Much of the damage resulting from the explosion is caused by flying glass.

Any potentially explosive mixture of gases is only dangerous within certain concentration limits. Thus, hydrogen and oxygen mixtures are only explosive between concentration limits of 4·65 and 93·9% $^{V}/_{V}$H. One of the main safeguards in handling gas mixtures is to keep the proportions outside the explosive limits.

OTHER SOURCES OF EXPLOSION

Another source of explosion is the violent decomposition of certain types of compound, liberating large volumes of gases at very high velocities; this can occur with such compounds as diazomethane, ozonides, peroxides, and heavy metal acetylides. It is unlikely that any of these explosive materials will be used by students in first-year work, but the possibility of forming peroxides must always be considered if hydrogen peroxide is being used.

Implosion

In normal laboratory work implosion is caused by the effect of atmospheric pressure on a closed, evacuated vessel that is either defective, or of unsuitable design.

Implosion of a flask or a bottle under vacuum is usually a consequence of using a container of unsuitable shape, in which the pressure difference results in excessive tensile strain on the glass. Glass containers intended to be used under vacuum should always be thick-walled and shaped in such a way that the pressure over the whole surface area is compressive rather than tensile. This effect is illustrated by the fact that a comparatively thin-walled spherical flask can be safely evacuated, because the pressure difference on the glass results in a compressive strain, but the walls need to be very much thicker to withstand the same excess pressure inside. The diagram also illustrates the reason for not using a flat-bottomed vessel under vacuum, unless it is especially designed for the purpose, in which case the wall thickness is much greater than would otherwise be needed.

It is highly dangerous to use ordinary bottles, especially large ones such as Winchester quarts, under vacuum, as these containers always have a flat or slightly concave (viewed from outside) base, and so when the internal pressure is reduced the glass is almost certain to fail sooner or later. This failure is even more likely to occur if the unsupported area is large, and the risk is further increased if the outer surface of the strained glass is scratched or otherwise damaged.

Most students have certain misconceptions about working with high vacuum equipment; they will evacuate a small, spherical flask on a water pump to give a pressure of about 12 mm/Hg, but hesitate to reduce the pressure further to, say, 0·01 mm. It should be appreciated, however, that under normal conditions the maximum possible pressure difference across the walls of an evacuated container can only be one atmosphere, and from the point of view of stress on the convex glass surface, the effect of reducing the pressure inside from 12 mm to 0·01 mm is not likely to add much to the risk of implosion. If there is any reason to doubt the ability of the container to withstand high vacuum conditions, it should not be used at any reduced pressure.

The risk of a glass desiccator collapsing under vacuum is always present, and it should be protected by a suitable metal cage or other container, to prevent glass flying if implosion should occur.

General Attitude to Hazards

It has to be accepted that work in a chemical laboratory involves some intrinsic hazards, but the real risk can be reduced to negligible proportions if a proper attitude of mind is developed. As in all other aspects of laboratory practice, the key to the problem is intelligent forethought; it is essential to develop a sense of anticipation about hazards. Before carrying out any procedure in the laboratory, the student should consider what kind of mishap could occur, how could it be prevented, and, if it does happen, what action will be required. If all of this thinking is done before the mishap—which is less likely to occur because of the forethought—no time is lost in deciding what to do, nor is the damage aggravated by hastily taking the wrong action.

Before starting any experiment the possibility of hazard or risk to people must be carefully considered. If necessary, others must be warned and the decision made in advance just where and how to move quickly if the need should arise. Once this habit of anticipatory thinking about hazards has been developed, the detail becomes self-evident. For example, if there is a fire risk, the decision should be made in advance about what kind of fire extinguisher is to be used and where to find it in the laboratory. If the fire risk is especially severe the extinguisher should be removed from its normal mounting and kept close at hand, immediately available for use. The extinguisher must, however, be returned, refilled if necessary, to its normal location as soon as possible.

When starting a reaction known to be vigorous, plans should be made in advance to deal with it should it need to be moderated, such as by cooling. This can often best be done by spraying the upper part of the reaction flask with water. Where a reflux container is used, it may be possible to do this effectively by using the water from the condenser outlet. If the reaction flask is being heated on a steam bath, turning off the steam will probably not produce a sufficiently rapid response, but this dousing treatment with the condenser water can be very effective. If this treatment is likely to be needed, the flask could be slightly raised away from the rings of the bath at the beginning of the experiment by placing a matchstick, for example, so as to leave a clearance between the flask and the hole in the ring. The cooling water can then run over the upper part of the flask, and not only cool it, but, in draining away, also cool the water in the bath.

Arising from this question of anticipation of hazards, certain other matters of general good practice should be noted. It is very bad practice to keep the hands in the pockets when working at the bench, since the ability to move quickly or to protect the face is severely impeded. It is also bad practice to sit in front of a flask containing boiling liquid. Should the flask break or a fire start, there is a serious risk of hot and possibly burning liquid cascading over the edge of the bench into the lap of a person sitting there. Stools cluttering up the floor-space also delay movement in emergencies.

CHAPTER 4

TECHNIQUES AND UNIT OPERATIONS FOR CARRYING OUT A REACTION

General Principles

STOICHIOMETRY

Most reactions in organic chemistry are characterized by being slow, reversible, and accompanied by side reactions. It is common practice to use one or more of the reactants in excess of the exact stoichiometric proportion, in order to:

(a) accelerate the reaction;
(b) influence the position of the final equilibrium, so that maximum yield of the required product is obtained;
(c) suppress side reactions;
(d) provide a solvent.

ORDER OF ADDITION OF REACTANTS

Since the yield of a reaction can be affected by having one or more reactants present in excess of the stoichiometric quantity, the order of addition can be very important, especially in minimizing side reactions. If one component of the reaction mixture is apt to undergo side reactions with the other starting materials, or with the reaction product, it should be added gradually to the rest of the mixture as the reaction proceeds (e.g. addition of halide in the Grignard reaction).

SOLVENT

The speed and direction of a reaction can be affected by the choice of solvent. Any reaction must be preceded by collision of the molecules of the reactants; the number of collisions will be much smaller if the reactants are not in the same phase, whether they be two immiscible liquids, a solid insoluble in a liquid, or two solids. In organic chemistry, a homogeneous reaction mixture is nearly always achieved by using a solvent; melting the reactants for this purpose is usually inconvenient and may not be feasible because many organic compounds tend to decompose when heated beyond their melting points.

The temperature at which the reaction is to be carried out can be controlled by using a solvent at its boiling point, which must not be so high as to cause decomposition of the reactants or reaction products, nor so low as to give too slow a rate of reaction. The solvent must not have any adverse chemical effects on the reaction system.

In some cases a solvent is added in order to retard reaction rate. In an exothermic reaction, the temperature of the mixture cannot rise above the boiling point of the solvent, and it can often be useful to add an inert diluent (with a suitably low boiling point) to reactants which are themselves miscible. In this case, the diluent helps to moderate the reaction by limiting the number of collisions between the reactive molecules, and by absorbing heat. It will be seen that heat produced in the reaction is first absorbed as latent heat in converting the liquid solvent to vapour. This latent heat is then removed by the condenser cooling water when the vapour is converted back to the liquid phase.

If one of the reactants dissolves the others in a mixture, and has a suitable boiling point, it may be permissible to use it as the solvent for the reaction. This can influence the yield of the final product (usually in a favourable direction), since a large excess of the solvent reactant must be added, and the position of a final equilibrium will be changed; for example, in the esterification of benzoic acid by methanol with sulphuric acid as a catalyst a large excess of the alcohol is used as a solvent and the reaction carried out by heating under reflux. To isolate the methyl benzoate from the mixture, at the end of the reaction, the excess methanol must be distilled off. As the methanol is removed, the reaction will tend to reverse itself; this principle applies to all equilibrium reactions where one of the reactants acts as solvent, and where the reaction product remains in solution. In such cases, the excess solvent must be distilled off as rapidly as possible.

The solvent may be chosen so that the main product of the reaction is not soluble in it. This makes the problem of isolating the product from the reaction mixture much easier; if the desired product is a solid, it may be recovered by filtration, while a liquid may be isolated by means of a separating funnel.

HETEROGENEOUS REACTIONS

It is not always possible to find a solvent which will provide a homogeneous reaction system, and alternative methods must be found to provide optimum reaction conditions. In the hydrolysis of an aromatic ester such as propyl naphthoate, for example, the reaction may be carried out in either aqueous acid or alkali. Propyl naphthoate is not very soluble in either medium, but the naphthoic acid which is formed by hydrolysis is soluble in aqueous alkali, but insoluble in acid. Since the reaction takes place at the surface of the particles of propyl naphthoate, if the hydrolysis is carried out in acid, an insoluble film of naphthoic acid will form on the outside of the particles of ester, the acid will fail to reach the inside of the solid, and the yield of naphthoic acid will be diminished. If, on the other hand, alkali is chosen as the hydrolysing agent, the naphthoic acid will dissolve in the alkali as it is formed, leaving the propyl naphthoate particles open to attack by the alkali. The reaction will then proceed to completion; when the hydrolysis is complete, the ester will have disappeared, leaving a homogeneous solution.

Stirring is always needed when the reaction mixture is not homogeneous, consisting of either two immiscible liquids, or a dense solid and a liquid. In the latter case, the stirring is especially important if heating is required (see p. 52).

Materials Handling

WEIGHTS AND VOLUMES

It is important to consider the precision required for measuring weights and volumes in preparative organic chemistry. Most student preparations require quantities of reactants and reaction products of the order of 50 to 0·5 g and quantities of solvents up to about 500 ml. Organic reactions rarely give 100 per cent yields, and side reactions are common, and an excess of one or more of the reactants may be used to push the equilibrium in the right direction. It is clear, therefore, that for most organic preparations, there is no advantage in weighing the reaction materials with anything like the same precision as that used for analytical purposes. For organic preparations, a weighing accuracy of about $\pm$ 0·5 per cent is usually acceptable; weighing to better than the nearest 10 mg is not likely to be needed and in many cases even the nearest 0·1 g is good enough. Therefore, in weighing out the materials for an organic preparation, or the reaction product obtained, it is neither necessary nor desirable to use a high-precision analytical balance designed to weigh to 0·1 mg or better.

SOLIDS

Dry solid reactants can be weighed on a piece of paper with a smooth glazed surface, creased down the middle to give a broad V shape. The use of filter paper can lead to significant losses in handling small quantities of fine solids, and fibres may become detached from the surface, introducing impurities. In any case the use of filter paper for weighing purposes is wasteful, and ordinary writing paper or paper from an obsolete catalogue of a laboratory equipment supplier is usually quite suitable. The weight of the paper must not be ignored, but if a two-pan balance is being used it is sufficient to counterbalance with a similar piece of paper. After weighing, the creased paper forms a chute and the solid is thus easily transferred into the reaction vessel. To avoid losses by spilling, the vessel should be standing on a large sheet of glazed paper so that any spilled material can be recovered. If the solid is wet, sticky or highly corrosive, it must be weighed on to a container which has previously been weighed, or "tared". It may be advantageous to weigh a reactant directly into the tared reaction vessel, but this can only be done for one substance. It is very bad practice to try to weigh out a second substance into a flask that already contains some other material; in the process of adjusting the weight it may be necessary to remove some material and if more than one substance is involved, it will be impossible to be sure that only the required one is being withdrawn.

Large crystals or lumps of solid reactants should be reduced to a finer texture by crushing or grinding in a pestle and mortar. This operation involves losses which may be quite considerable, and must therefore be done *before* weighing a reactant, so that there is no doubt about the actual weight used for the reaction. By the same reasoning, if a reaction *product* is to be crushed or ground, it should be done *after* the weight (and hence, the yield) has been determined.

If very small quantities of a solid reactant are required and if a solvent is to be used in the reaction, it is sometimes convenient to make up a stock solution of the solid of a known concentration (e.g. 10 per cent w/v) in that solvent, and then transfer to the reaction mixture a measured volume of the solution calculated to contain the required amount of solid. When calculating the total solvent required for the reaction mixture, allowance should be made for the volume contained in the added solution. If one of the solids is difficult to handle, because of its sticky texture for example, it may be convenient to dissolve one or more of the weighed reactants in the reaction solvent using a suitable vessel such as a beaker or conical flask and then to transfer the solution to the reaction flask.

When transferring solids to a reaction flask it is important to see that none remains clinging to the neck, especially if this is a ground-glass socket into which a cone or stopper is later to be inserted. Surplus solid clinging in this way may be pushed into the flask with a small spatula, or if the reaction mixture includes a solvent, some may be reserved to be added last, to wash any such residual solid down into the flask. It is often possible to remove such surplus material by carefully wiping with a cloth or duster, but this results in a small loss of solid from the reaction mixture.

The method of addition of materials to a flask should always take account of special hazards. Thus, a flask may easily be broken by dropping large sticks of sodium hydroxide into it; this type of addition should be made with the flask so positioned that the neck is almost horizontal and the sticks slide into it with a minimum distance to fall before striking the glass. In this particular case, the problem can be eased by using pellets rather than sticks of sodium hydroxide. The risk of breakage when adding a solid to a flask can be further reduced by having some liquid in it to act as a cushion, but then if the liquid is corrosive or otherwise harmful, it is especially important to prevent splashing. This risk is at its worst when the neck is pointing vertically upwards, and it is therefore more than ever important to have it nearly horizontal when adding the solid.

LIQUIDS

All liquids except mercury can be assumed to wet glass, and so, however carefully a liquid is transferred from one vessel to another, some will be left behind in the form of a film, wetting the inner surface of the container. The amount of these drainage losses will depend upon two factors, the area of surface wetted, and the viscosity of the liquid; they do not depend directly on the quantity of liquid in the flask. Therefore the absolute loss for any one flask being emptied is virtually constant, although the percentage loss is inversely proportional to the total weight of liquid. The percentage loss in transferring a small quantity of liquid from a large container is therefore especially high. When handling liquids the number of transfers should obviously be minimized and the smallest possible container chosen. It is sometimes possible to eliminate drainage losses by using a suitable solvent to wash out liquid from the original container into the receiver. In the case of a reactant, the solvent which will be used in the reaction is the obvious choice. When a reaction product is being transferred, the solvent chosen should be easily removable. Ether is commonly used because of its good solvent properties, its low viscosity (i.e. it drains well) and the ease with which it can be removed by subsequent evaporation or distillation from the dissolved material. It must be remembered, however, that ether will dissolve about 1 per cent of water, so this solvent must be dry when used to transfer a dried liquid.

Liquid reactants can be measured by weighing into a tared container, which may be the reaction vessel. If more than one liquid is involved, a clean tared container must be used for each weighing. It is bad practice to weigh a liquid into a container that already contains some other substance because it is impossible to remove any excess liquid inadvertently added, without also taking out some of the material originally present. Liquids are commonly measured by volume, and ordinary (grade B) measuring cylinders are quite adequate, especially if the precision of measurement need not be high, such as when dealing with a reaction solvent, or one of the reactants which is being used in a large excess.

When using a graduated cylinder to measure a viscous liquid it must be remembered that the quantity used in wetting the glass surface can be surprisingly large, and that serious errors in measuring the volume can occur. The drainage error will be especially serious when measuring small quantities. If, for example, 1 ml of concentrated sulphuric acid is required, and that quantity is simply poured into a 10 ml cylinder and then into the reaction mixture, considerable error will arise because a very large area of the glass surface of the cylinder will be wetted by the liquid when it is poured out, and the amount added to the reaction mixture will be considerably less than the intended volume. This can be demonstrated by leaving the "empty" cylinder to stand for about 15 minutes after removal of the contents, when it will be seen that a substantial proportion of the original 1 ml is left in the cylinder. There are various ways of dealing with this problem: in the particular case of the sulphuric acid it is best to use a small measuring cylinder, which will be reasonably full when used for measuring the desired quantity. Further, if it is necessary to measure only 1 ml, it is best to add about 2 ml of sulphuric acid to the cylinder initially. The exact volume in the cylinder is then noted and an estimated 1 ml is transferred dropwise from this to the reaction mixture. The cylinder is then allowed to drain back and the volume reading noted again. The volume difference will represent the amount removed from the cylinder and with a little practice this method can be used quite successfully. Even if the amount transferred from the cylinder differs considerably from the intended 1 ml, at least the volume will be known with reasonable accuracy.

In many cases, drainage errors in measuring viscous liquids can easily be minimized if the reaction mixture includes a solvent. In that case, the desired volume of the viscous material can be measured out in the usual way in the cylinder, which is then emptied into the reaction flask and the residue washed from the cylinder into the flask with some of the solvent. Alternatively it is sometimes convenient to prepare a solution of the viscous material in the reaction solvent to a known concentration and an appropriate volume may be transferred to the reaction mixture (sulphuric acid may not always be suited to this technique). The quantity of solvent present in the solution must be allowed for when subsequently adding the rest of the solvent.

The effect of drainage errors should be remembered when collecting fractions from a distillation. The receivers must be tared before use so that the weight of liquid collected in them can be determined directly from the gross weight without having to transfer to another tared vessel. If, inadvertently, the weight of receiver is unknown, the best procedure is to weigh it with the the collected liquid which ist hen drained into another vessel. The original receiver can then be thoroughly cleaned and dried, and finally weighed.

GASES

Students are advised to consult the laboratory supervisor whenever gases are required for a reaction. Hydrogen, nitrogen and oxygen are supplied in steel cylinders at pressures up to 150 atmospheres, while the more compressible materials such as ammonia, chlorine and carbon dioxide are supplied in similar cylinders at rather lower pressures.

All cylinders are fitted with a valve, but in the case of cylinders where the contents are at high pressures, it is very dangerous indeed to try to remove gas without fitting a special regulating valve which will automatically reduce the pressure of emerging gases to a pre-set and suitably low level. Expert guidance should be obtained in the technique of fitting or removing regulator valves.

Compressible gases such as ammonia are frequently supplied in cylinders which do not normally require a pressure regulator but special valve fittings are still necessary and expert advice on these should be obtained.

ANHYDROUS CONDITIONS

Many organic reactions are adversely affected by water and in some preparations special precautions must be taken to exclude it. Because of its low molecular weight, a small quantity of water can cause considerable chemical damage. Consider, for example, an acid chloride with a molecular weight of 180. This may be hydrolysed by water in accordance with the equation:

$$\underset{180}{\mathrm{R{-}COCl}} + \underset{18}{\mathrm{H{\cdot}OH}} = \mathrm{R{\cdot}COOH} + \mathrm{HCl}$$

from which it can be seen that only 18 g of water is capable (theoretically) of hydrolysing 180 g of the chloride. Suppose now that for purposes of some reaction (other than hydrolysis) 18 g of the acid chloride were dissolved in ether (100 ml) containing 0·5 per cent of water. This quantity of moist ether, containing 0·5 g of water, is capable of hydrolysing ten times that weight (i.e. 5 g) of the acid chloride under discussion, so that some 28 per cent of the amount of the latter originally dissolved in the ether is likely to be destroyed.

To achieve anhydrous conditions, the chemicals as well as the apparatus must be dried, and atmospheric water vapour must be prevented from entering the apparatus. This can be done with stoppers, or with a guard tube filled with a suitable desiccant such as anhydrous calcium chloride.

Filling Guard Tubes

A loose plug of glass wool is first pushed into the bulbous part of the tube, which should be no more than half filled. The granular drying agent (calcium chloride, soda lime or silica gel) is then added, so as to fill the tube to within about 2 cm of the open neck. The solid is retained by another loose plug of glass wool (Fig. 4.1).

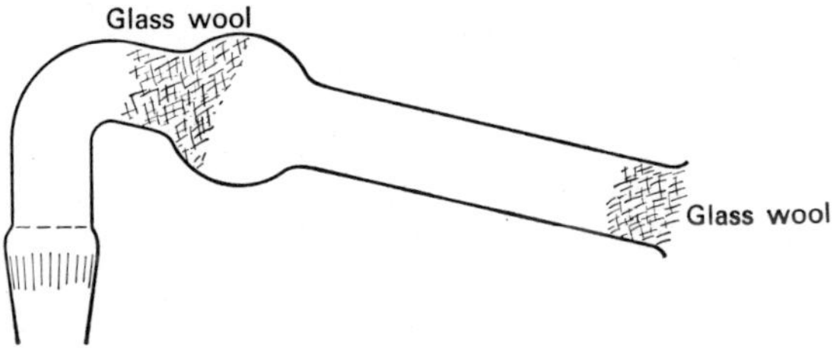

FIG. 4.1. Drying tube.

Cotton wool is sometimes used instead of glass wool, but it is not very satisfactory, because it packs down too much. The guard tube should not be mounted vertically on top of a reflux condenser, especially if the reaction mixture contains alcohol. Small amounts of vapour from the condenser may enter the guard tube, dissolving the desiccant, and contaminating the reaction mixture as it drains back. The guard tube should always be fitted at an angle so that any liquid accumulating in it drains away from the reaction mixture.

Drying Reactants

Drying of reactants is discussed on p. 157. Many solvents, such as ethanol, are hygroscopic and must be kept in well-stoppered bottles. An important exception is ether drying over sodium; this must be stored in a vented container to permit escape of hydrogen produced by reaction between water and sodium; a calcium chloride tube is usually sufficient.

Sodium reacts immediately with water, and is therefore kept under light hydrocarbon oil. The metal should only be removed just before use. If the oil does not interfere with the reaction, no attempt should be made to remove it; if the oil must be removed this can be done by dabbing with filter paper after which the sodium is finally cleaned by dipping it momentarily into absolute ethanol. If some other alcohol is being used in the reaction, a sample of it is used to free the sodium of last traces of oil.

Assembly of Apparatus

SUPPORTS

The more supports used, the greater is the risk of glassware being subjected to bending strain resulting from misalignment. Glass has great strength when under compression, but it is relatively weak under tension, and badly aligned supports are the cause of virtually all glassware breakages from causes other than dropping. The minimum number of retort stands should be used, although if too many clamps are clustered on to one retort stand it may be difficult to reach any one of them if part of the apparatus has to be changed or adjusted during the course of the experiment. A technique of assembling the apparatus that minimizes risk of strain and breakage is illustrated in Fig. 4.2. The base-plate of the retort should be immediately below the main weight to be supported, and the boss-head positioned on the retort stand with the shaft of the clamp secured loosely on to it (Fig. 4.2(a)). Only one of the jaws of a clamp moves when it is being tightened round a piece of glassware (Fig. 4.2(b)); with the jaws of the clamp wide open, it is adjusted to bring the fixed portion comfortably round the glass to be secured. The rod of the clamp is then secured into the boss-head, and finally the latter is tightened on to the retort stand, taking care that the position of the clamp relative to the

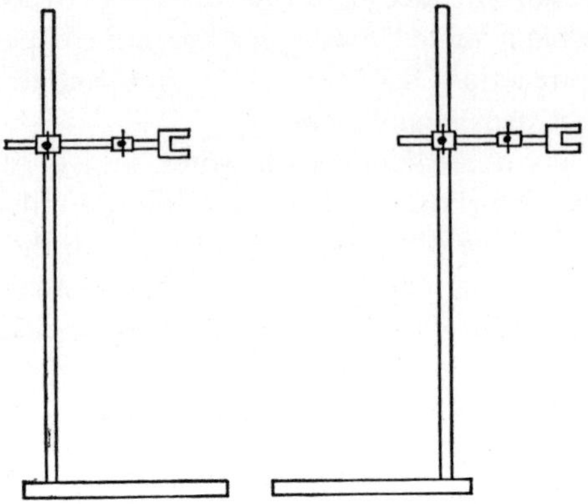

FIG. 4.2. Use of retort stand.

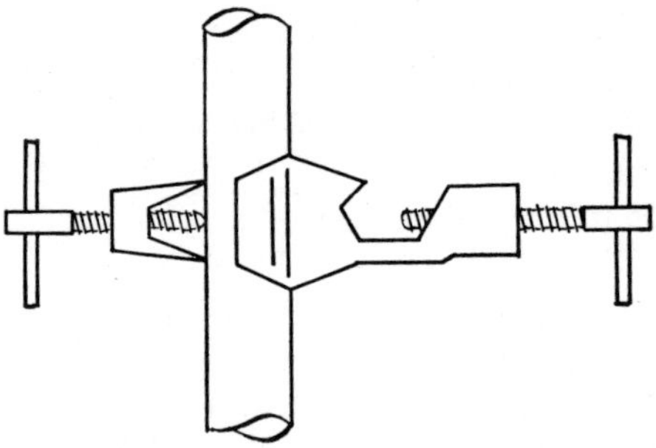

FIG. 4.2(a). Attachment of boss to retort stand.

glassware is not changed during these operations. Then the clamp screw is carefully tightened, so as to bring the moving jaw into position to secure the glass.

When setting up an apparatus for operations such as reflux or distillation, the heater system will often provide a sufficient support for the boiler flask, and clamping may be unnecessary, especially if an electric heating mantle or steam bath is used. In many cases

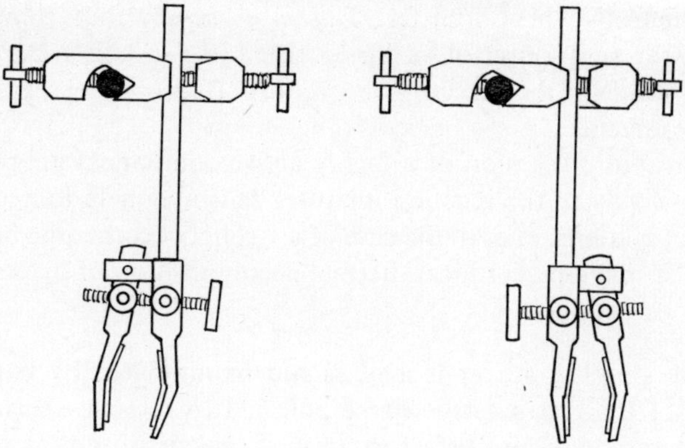

FIG. 4.2(b). Assembly of clamp and boss.

a gauze or tripod will provide adequate support for a small flask, and it should only be necessary to clamp the remaining part of the assembly, such as a reflux condenser: if the reflux condenser is to be removed before the end of the experiment, it will be desirable to clamp the neck of the flask as well. In a reflux assembly it is *not* satisfactory to clamp the boiler flask and leave the condenser unsupported, especially if a water-cooled condenser is used, because the sideways pull of the water tubes causes considerable strain and may lead to breakage. When using an oil bath as the heating medium, a clamp should always be used to support the boiler flask.

With a system working under vacuum, the boiler flask will be supported by the atmospheric pressure, but it must be remembered that when the apparatus is opened to the atmosphere further support may be needed.

AGITATION

A reaction mixture may be stirred to achieve the following:

(a) maximum contact of the reactants, especially if the mixture is heterogeneous (e.g. solid in liquid or two immiscible liquids);

(b) effective heat transfer. This is especially important if the external source of heat is localized (e.g. a bunsen burner), or if one of the reactants or products is a solid denser than the solvent;

(c) rapid dispersion of a highly active component being added slowly to the reaction mixture. This may help to avoid side reactions, and, in the case of a strongly exothermic reaction, it may prevent local thermal decomposition of the reactants or product.

The simplest stirrer is a glass rod manipulated by hand; the ends of the rod must be flame-polished to remove sharp edges. Glass tubing is unsatisfactory even if sealed to prevent material entering, because it is not strong enough. An exception is the use of a test tube as a stirrer for a small quantity of a stiff pasty suspension.

There are many types of mechanically driven stirrer, for use in open or closed systems.

Unsealed stirrers are used with open-top vessels, and although no seal is required, a guide tube is usually necessary to provide steady running conditions.

For mobile liquids, the stirrer may simply consist of a piece of bent glass rod (Fig. 4.3(a)). To improve efficiency, the short, bent limb of the stirrer may be flattened by heating it to red heat in a flame and then squeezing it with a pair of tongs to give a "hockey-stick" shape (Fig. 4.3(b)). This type of stirrer is run at speeds up to about 300 r.p.m. and may be used for liquids with viscosity comparable to that of glycerol.

A stirrer guide, consisting of a loose-fitting sleeve of glass tubing around the long limb of the stirrer, is clamped to provide a steady bearing (Fig. 4.4). The stirrer is connected to the shaft of the driving motor by means of a short length of rubber pressure tubing, and adjusted so that there is a small clearance (about 5 mm) between the end of the stirrer and the driving shaft. This

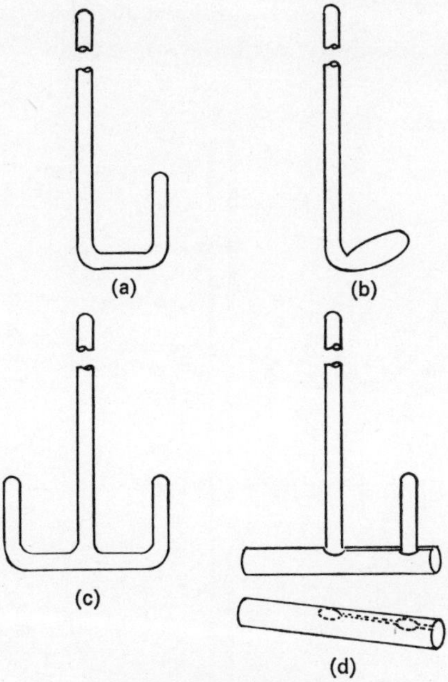

FIG. 4.3. Simple stirrers.

forms a very effective flexible coupling and will accommodate considerable misalignment between the motor and the stirrer shaft. To guard against the possibility of the stirrer shaft becoming detached from its coupling and dropping through the bottom of the flask or beaker, a thin tightly fitting retaining collar, cut from a piece of pressure tubing, is fitted over the shaft so that it just clears the top of the guide when assembled for normal use.

If the mixture to be stirred is a thick, pasty slurry, a small stirrer simply cuts a hole in the centre, leaving the main part of the suspension undisturbed. An effective stirrer must have a large sweep and rotate slowly (50–100 r.p.m.). A stirrer of this type (Fig. 4.3(c)) can easily be fitted into a beaker, or wide-mouthed flask, but for a narrow-necked flask, a device with a flexible link must be used. An ordinary "hockey-stick" or similar stirrer fitted with a length of stiff, plastic tubing which has been split longitudinally is a good improvization in some cases (Fig. 4.3(d)).

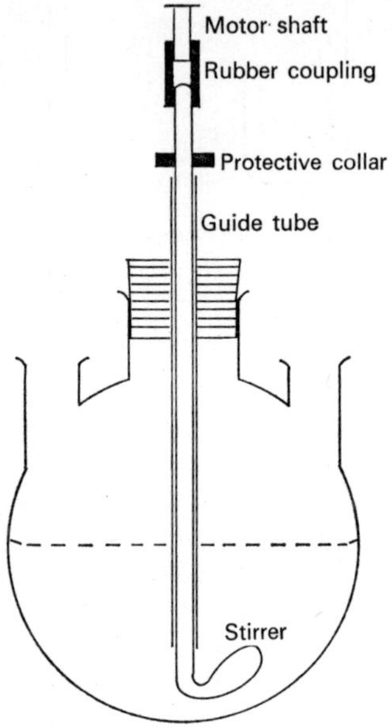

FIG. 4.4. Simple sealed stirrer assembly.

Sealed stirrers are used when the escape of vapour or the ingress of air must be prevented.

For small-scale work, a magnetically driven stirrer is satisfactory. The stirrer itself is a small cylindrical magnet, coated with a chemically resistant material (glass or plastic) which is placed in the mixture to be stirred. Under the reaction vessel is placed a unit consisting of a powerful magnet rotated by a variable-speed electric motor. The interaction of the two magnetic fields causes the stirrer to rotate when the driver magnet is set in motion. It is usually necessary to place the driver magnet quite close to the bottom of the flask for the system to give satisfactory results, so it cannot be used when a heating bath is required; it is also unsatisfactory for stirring large quantities of materials. When magnetic stirrers are not suitable, the simplest type of seal for a mechanically driven stirrer is achieved by making the guide long enough for its lower end to remain below the surface of the liquid in the flask, even in presence of the vortex produced by the rotation (Fig. 4.4). The main limitation of this type of seal is that it is only effective

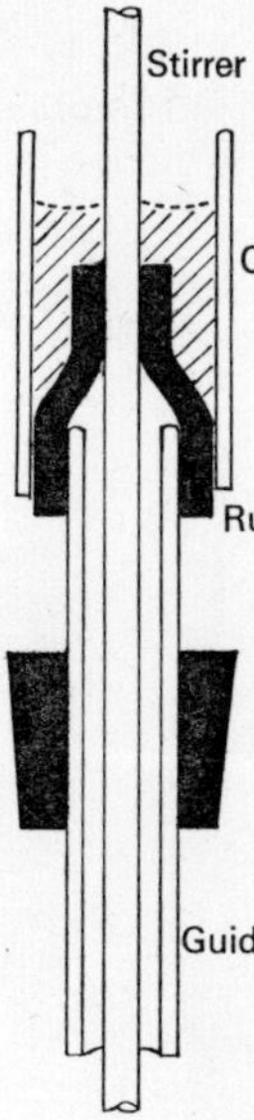

FIG. 4.5. Sealed stirrer assembly.

when the pressure within the flask does not differ from atmospheric by more than the head of liquid above the base of the stirrer guide. This stirrer is excellent for use in a reaction mixture being heated under reflux at atmospheric pressure.

If air or atmospheric moisture must be prevented from reaching the reaction mixture, this simple stirrer guide seal may not be satisfactory. Another simple seal can be formed between the stirrer guide and the rotating shaft of the stirrer, by fitting a piece of flexible tubing very firmly on to the stirrer guide, but less tightly on to the stirrer shaft (Fig. 4.5). If the area between the tubing and the stirrer shaft is lubricated with liquid paraffin, glycerol or silicone oil depending on the reaction mixture and the operating conditions, the stirrer will continue to rotate quite smoothly. Constant lubrication should be provided for a long period of operation, from a reservoir containing the lubricant. This type of seal works quite well even when there is a substantial pressure difference between the inside and outside of the flask, but is most suitable when the pressure outside is higher, producing a self-sealing action.

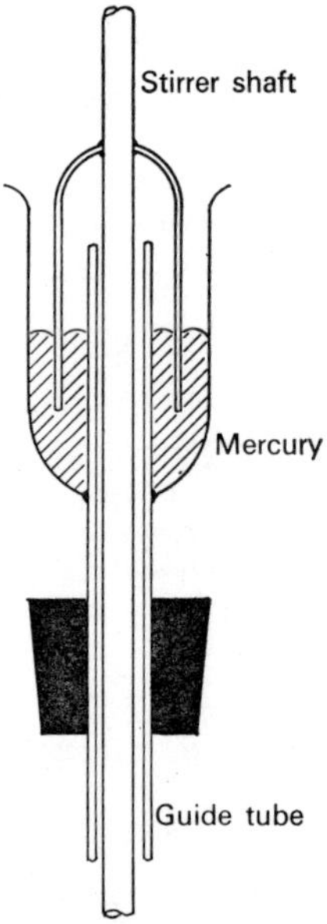

FIG. 4.6. Mercury sealed stirrer.

The mercury seal (Fig. 4.6) works well unless the pressure inside the flask is considerably lower than atmospheric, when mercury may be sucked inside the flask. Mercury is chosen as the liquid in the seal in spite of its toxicity, because of its high density. It is now possible to purchase stirrer gland fittings in which the seal is provided by a chemically resistant rubber ring compressed round the stirrer shaft by a threaded ring.

In all cases the efficiency of a stirrer is greatly improved by assembling it so that the stirrer shaft is offset with respect to the axis of symmetry of the flask to which it is fitted. This produces turbulence, and much more effective mixing than a stirrer symmetrically aligned with the flask. The effect is most obvious when a high-speed stirrer of small sweep is used.

Heat Transfer

The process of heat transfer is used in organic chemistry to achieve one or more of the following effects:

(a) change of reaction velocity,
(b) change in solubility,
(c) change of state (i.e. melting, vaporizing, condensing).

Although the practical techniques of heating and cooling are described separately, it should be appreciated that both are examples of heat transfer, and the same general principles are applicable to both.

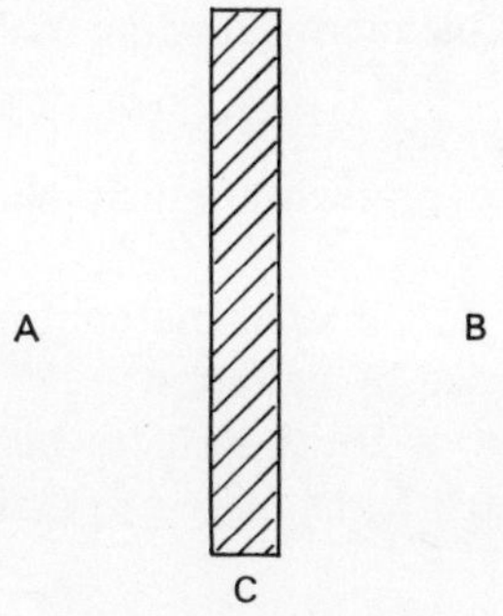

FIG. 4.7. Heat transfer.

Consider the system (Fig. 4.7) of two liquids, *A* and *B*, separated by a solid, non-porous partition *C*. The factors governing heat transfer from *A* to *B* are:

(1) *A* must be at a higher temperature than *B*, and the rate of transfer will increase as the temperature difference increases.

(2) The rate of heat transfer will also depend on the nature of the barrier *C*, and the factors here include:

(a) the thermal conductivity of the material;
(b) its thickness;
(c) the area of *C* in contact with *A* and *B*;
(d) the smoothness of the surface.

(3) Because heat must pass through the bulk of *A* to reach *C*, and similarly, heat must be conducted away through the bulk of *B*, the thermal conductivities of *A* and *B* will also affect the rate of heat transfer.

(4) Agitation of *A* and *B* will increase the rate of heat transfer, especially if the thermal conductivities of *A* and *B* are low, by minimizing the effects of the static films of liquid at the interfaces between *C* and the liquids *A* and *B*.

In practice, some of these factors are not easily changed; *C* is usually made of glass, and its thermal conductivity cannot be altered, although apparatus may be selected for minimum thickness and maximum surface area.

The easily controllable factors leading to maximum rate of heat transfer are:

(a) the temperature gradient, which should be as large as possible;
(b) static layers of hot or cold material at the glass interfaces must be eliminated by agitation;
(c) the contact area of the interfaces between the hotter and colder media must be as large as possible.

HEATING

The temperature of a liquid being heated in a glass container will lag at least 10° behind the outer surface temperature. If the liquid is being boiled, the latent heat of vaporization required will make greater demands on the heating system, and the outer surface temperature may need to be 30–40°C above the boiling point of the liquid.

Heating Baths

Hot water or steam baths are useful for obtaining temperature approaching 100°C. Because the rate of heat transfer is proportional to the surface area being heated, higher temperatures are attained more rapidly by using a round-bottomed flask surrounded by steam than a flat-bottomed flask on top of a bath. To limit the rate of input, e.g. when boiling a volatile liquid, the heated area may be reduced by using smaller rings, or the steam flow may be cut back.

Boiling water baths are usually provided with a constant-level device to keep them from boiling dry. Water should be fed into this device at a rate sufficient to give a small amount of overflow.

For small-scale preparations it is often convenient to use a beaker or a metal dish containing water or other suitable liquid, and heated on an electric hot plate. For temperatures above 100°C, oil is used as the heating medium. Ordinary liquid paraffin can safely be heated to about 200°C, but prolonged use at these higher temperatures causes decomposition (cracking) accompanied by darkening of the colour and a decrease in the flash point. This constitutes a fire hazard, and the oil should be discarded when it is pale brown. For prolonged use at, or above 200°C, silicone oils are usually satisfactory, especially for small-scale operations, but for large-scale working it is customary now to use electric heating mantles. One of the main advantages of a liquid bath is that its temperature is easily measured, and a thermometer should always be used with this type of heating system.

Electrical Equipment

Electric hot plates are useful for heating flat-bottomed containers such as beakers, but are not suitable for vessels of other shapes, such as flasks, for which special heating mantles are now commonly used.

Electric heating mantles consist of a flexible glass fabric, shaped to accommodate a glass flask; an electric heating element is sewn or woven into the fabric. Different sizes are available, and the flexibility of the fabric makes it possible for the attached heating element to come into close contact with the flask to be heated. The flask must always be the right size for the mantle, and must contain some liquid, so that heat is conducted away from the mantle at a reasonable rate. Failure to observe this precaution may cause the element to burn out, damaging the glass-fibre fabric, and possibly the flask. The electricity supply must always be switched off when the mantle is empty. The mantles normally available are satisfactory for temperatures up to about 200°C; above this temperature, glass fibres become brittle, and may fuse. Special mantles with the fabric made of fused quartz fibres are used for higher temperatures. Flexible tapes fitted with heating elements can be used to provide localized heating, around a glass fractionating column, for example. Conical mantles may be used to heat funnels during hot filtration procedures.

Temperature Control

Electric hot-plates are commonly fitted with two or more heating elements, and a selector switch is provided to control the number of elements in circuit. Further control is provided by having these elements of different power ratings. Heating circuits can be provided with a device usually described as an energy controller (e.g. "Simmerstat") which automatically switches the current on and off repeatedly over a time cycle of about a minute. This device is adjustable so that the proportion of the time cycle in which the current is on may be varied. Because of the supply of power on an "all-or-nothing" basis, the apparatus being heated may show severe temperature fluctuations. With equipment such as electric hotplates or heating mantles for flasks containing relatively large amounts of liquid, the heat capacity of the metal or liquid is usually quite sufficient to smooth out these temperature fluctuations; where the heat capacity of the system is low, however, or where very precise temperature control is required, the current, suitably adjusted, must be supplied continuously; this can be done with a rheostat or variable transformer. The rheostat is cheap and relatively easily available, but suffers from the disadvantage that it is electrically inefficient, and may dissipate a large amount of heat. The variable transformer is excellent; it is efficient but rather expensive.

COOLING

The conventional cooling media used in organic chemistry are (a) tap water, (b) ice and water, (c) ice and salt, (d) solid carbon dioxide, and (e) liquid nitrogen or liquid air.

Condensers

Tap water is the usual cooling medium in condensers, although alcohol cooled by solid carbon dioxide can be used for difficult conditions.

The rate of heat exchange depends on both the temperature gradient between the hot vapour entering the condenser and the cooling medium in the jacket, and the area of contact between the hot vapour and the cooled surface. If water is the cooling medium for a low boiling-point distillate the area of the cooling surface must be as large as possible. An ordinary Liebig condenser might have to be unduly long to meet this requirement; various other types of condenser are available, and the general aim is to provide the maximum cooling surface within the minimum outside dimensions. For the maximum rate of cooling, the velocity of the cooling water flow through the condenser jacket should be as high as possible, and to achieve this, the clearance between the inner and outer tubes of a condenser must be minimal. The main points to remember in using a condenser are as follows:

(1) The condenser must always be at as sharp an angle to the

horizontal as possible, to provide maximum drainage of the condensate.

(2) The cooling water must always be led into the lower end of the jacket. If the water enters at the upper end, it will drain out so that only part of the area of the cooling surface is exposed to the water, and the working efficiency of the condenser is reduced.

(3) The water flow through the jacket should be adjusted so that the coolant leaving the condenser is at a temperature a few degrees higher than that entering it. When liquids of higher boiling point are being condensed, the outlet temperature may be allowed to rise higher. It is wasteful, and hence bad practice, to have the water flow greater than is necessary. The rubber inlet tube connecting the condenser to the water supply may be subjected to relatively high pressure; thick-walled (pressure) tubing should be used for the inlet, and it should be wired to the tap and to the condenser. In order to prevent the outlet from moving out of the drain, and possibly causing a flood, the rubber tube from the condenser should terminate in a length of glass tubing.

Water cooling of a substance in a flask or beaker is much more efficient if the water is kept moving, either by using a continuous stream, or by stirring the contents of the cooling bath. The material being cooled must also be stirred as much as possible, because heat transfer is seriously impeded by a static layer in contact with the cooling surface.

The contents of a conical flask may be cooled by covering the mouth with a small inverted beaker, and allowing the cooling water to flow over the top and in a cascade down the sides (Fig. 5.21).

Ice

Ice is often used for cooling mixtures. It is most effective if mixed with water, to maximize the area of contact between the cooling medium and the surface to be cooled. There should be sufficient iced water to immerse the beaker or flask above the level of the contents, and agitation of the coolant and the liquid to be cooled adds to efficiency. If a solid is deposited from a mixture being cooled in a bath, agitation is especially important to prevent the accumulation of a crust of crystals on the inner surface of the container; such a deposit will seriously impede the cooling process. When the addition of water to the liquid to be cooled is not objectionable, crushed ice can be added directly; this gives the maximum rate of cooling, but sometimes, to avoid contamination, the ice must be made from distilled water.

Ice and Salt

Ice and common salt is used for temperatures down to about −5°C. The ice should be finely crushed, and the container for the cooling mixture filled alternately with layers of ice and salt, so that there is the best possible mixing of the two materials.

Solid Carbon Dioxide

Solid carbon dioxide is an excellent cooling medium, and can be used to provide temperatures down to −78°C. It is not very satisfactory as a cooling medium when used dry, and is usually suspended in a water-miscible organic solvent. Methanol, ethanol, acetone and isopropyl alcohol are all satisfactory, but water is not suitable. Where practicable it is an economy to use recovered solvent for this purpose. If solid carbon dioxide is added to one of these organic solvents at room temperature, gas is liberated very vigorously, and the violent frothing will lead to loss of solvent. It is much more satisfactory to put the solid carbon dioxide, broken into small pieces, or powdered, into the container, and then pour the solvent slowly onto it.

Solid carbon dioxide can cause severe skin burns if handled carelessly, and prolonged contact with the skin must be avoided. The solid may be crushed by enclosing it in a strong cloth bag and hammering or beating it. Gloves should be worn, or the solid should be picked up with tongs or a scoop.

Liquid Nitrogen or Air

Liquid nitrogen or air are mainly used in traps for high vacuum systems. Precautions must be taken to prevent splashes on to the skin, or more particularly into the eyes, because of the extremely low temperatures (about −200°C).

CHAPTER 5

TECHNIQUES AND UNIT OPERATIONS FOR ISOLATING AND PURIFYING A REACTION PRODUCT

Distillation

The distillation process comprises the following stages:

(1) Heating a liquid to its boiling point and then converting the liquid to vapour.

(2) Cooling the vapour to convert it back to the liquid phase.

(3) Collecting the condensed vapour.

Before discussing the practical techniques involved in this type of operation it is necessary to consider some of the elementary principles in some detail.

Heating and Vaporizing

The volatility of many organic substances makes it possible to separate the individual components of mixtures by distillation, taking advantage of the fact that the boiling points of the components may be widely different. The boiling point of any substance is defined as the temperature at which its vapour pressure is equal to the pressure of its surroundings. In an homologous series, the boiling point usually tends to increase with molecular weight; other factors, such as hydrogen-bonding, also affect boiling point.

Effect of High Temperatures

The application of the technique of distillation is sometimes limited by the fact that heating above about 150°C causes many organic materials to decompose. Heating is a process of applying energy to a system; by raising the temperature beyond about 150°C the energy input approaches the bond energy levels within the organic molecule so that rupture can occur, producing free radicles or other species as precursors of polymerization or other reactions. Under these conditions, also, oxygen from the air is likely to react with the organic material.

Reduced Pressure—Effect on Boiling Point

By reducing the ambient pressure, the saturation vapour pressure of a substance is reached at a lower temperature—that is, the boiling point is reduced, and thus, decomposition by overheating is avoided.

It is usual to work in one of the following pressure ranges:

Atmospheric pressure: about 760 mm Hg
Water pump vacuum: 11–25 mm Hg
High vacuum (mechanical oil pump): 0·1–0·01 mm Hg
Extra high vacuum (vapour diffusion pump): about 1 micron or lower (10^{-3}–10^{-6}mm Hg).

It may seem that there are very wide steps in between one type of pumping system and the next, but they are quite sufficient for the vast majority of volatile organic compounds because vapour pressure is a logarithmic function of temperature. This means that for most organic substances the boiling point falls about 10°C each time the pressure is halved. Therefore if the boiling point is too high at 760 mm (say 220°C) there is not much to be gained by reducing the pressure to 380 mm because the boiling point will still be about 200°C; it will only be possible to obtain a useful reduction in the boiling point by taking the pressure down to about 10–20 mm. Even that may not be enough and in that case it will be necessary to go down to about 0·1 mm (oil pump) or even lower (diffusion pump).

Techniques for Distillation at Atmospheric Pressure

These can be discussed under the following headings:

1. Simple distillation,
2. Steam distillation,
3. Fractional distillation.

SIMPLE DISTILLATION

Solvent Removal and Stripping

A very common application of this is the isolation of a material obtained by solvent extraction from an aqueous solution or suspension. The solvent, which is subsequently removed by distillation, is assumed here to have a much lower boiling point than the material dissolved in it. The following discussion will be related to the distillation of ether, but the principles will apply equally to other solvents, except that in some cases the fire hazard may be diminished or absent.

The apparatus and techniques used for solvent stripping must take account of the following factors.

Heating. The solvent being removed has been selected for its low boiling point, so for most distillations a boiling water bath is adequate to produce a temperature on the outer surface of the boiler flask about 30–50°C above the boiling point of the liquid being distilled. Precautions must be taken to avoid excessively rapid boiling which may cause some of the dissolved material to be carried over with the ether; an anti-splash head (Fig. 5.1) will help to prevent this. If a steam bath is used, the rate of boiling may be restricted either by having a small surface area of the flask exposed to the steam (only a few rings are removed from the lid of the bath) or by raising the flask so that it does not actually rest on the rings. The former method is better. There is no justification whatever for using a bare flame when distilling ether, and it is also dangerous to try to do the distillation of the ether with the boiler flask immersed in a beaker of water heated with a Bunsen burner. If a steam bath is not available, the water for the bath should be heated over the Bunsen well away from the distillation apparatus. This procedure is rather primitive and time-consuming, but it is safe.

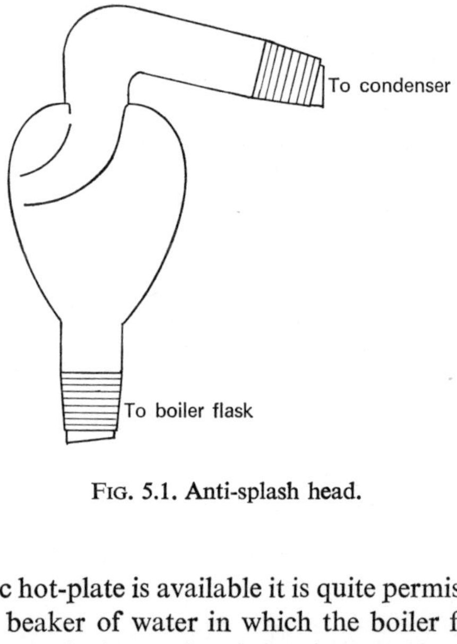

FIG. 5.1. Anti-splash head.

If an electric hot-plate is available it is quite permissible to use it for heating a beaker of water in which the boiler flask is partly immersed. Alternatively, an electric heating mantle with a suitable controller system may be used but temperature regulation is less satisfactory with this form of heater.

One of the main advantages of a water bath is that regardless of the rate of heating the temperature cannot exceed 100°C and there is usually no risk of damage to the residue left after all the ether has been removed. For higher temperatures, an oil bath may be used; this should always be provided with a thermometer, which may also be used to stir the oil to promote even temperature distribution.

Bumping. In all distillation operations there must be smooth conversion of the liquid to the vapour phase without any significant degree of disequilibrium associated with super-heating and bumping. Most liquids have relatively poor thermal conductivity; this allows a situation to develop where heating at the bottom of the flask results in local boiling while the temperature of the main bulk of liquid is appreciably lower. Under these conditions vapour can accumulate in a pocket below the surface of the liquid until the pressure developed locally is sufficient to release it with explosive violence. The repeated occurrence of these events—build-up of pressure, followed by explosive release—is known as bumping. The usual cure for this is anti-bumping stone, which is simply some very highly porous ceramic material such as unglazed earthenware and is used in the form of small pieces about the size of a rice grain. The stone must be unglazed—pieces of glazed crucible are not suitable, nor are pieces of fused silica because this, like glass, is not porous. Pumice is sometimes used, but its low density sometimes causes it to float; for successful operation it is essential that the stone should rest on the bottom of the flask. For most purposes one or two small pieces of the anti-bumping material are quite sufficient; it is bad practice to use more because no further improvement in boiling is achieved and it leads to losses of the product by occlusion of liquid in the pores.

The anti-bumping stone must always be added before heating is started. If added to a super-heated liquid it will cause very vigorous boiling and frothing that can lead to loss of material or even result in a fire. If the addition of anti-bumping stone has been overlooked when heating is started, the hot liquid must be allowed to cool to below its boiling point before it is added. Anti-bumping stone used in a boiling liquid loses its efficiency once the liquid has been cooled, so if it has been necessary to interrupt the boiling, it is essential to cool the liquid slightly and add more stone before resuming the distillation.

An alternative to anti-bumping stones is a wooden splint long enough for one end to be well above the liquid level when resting in a near-vertical position with the other end on the bottom of the flask. An objection to the use of wood is that there is the possibility of leaching from it by the action of the solvent, with consequent contamination of the liquid. To avoid this, it is possible to use a glass device made from thin-walled tubing, and shaped as in Fig. 5.2(a) or (b). This also is supported in the flask in a near-vertical position, with the bulb resting on the bottom. Although these devices sometimes have advantages in the distillation of very large volumes, or for liquids of very high density, they are of limited value, and for nearly all student work, the use of porous pot or similar anti-bumping granules is recommended.

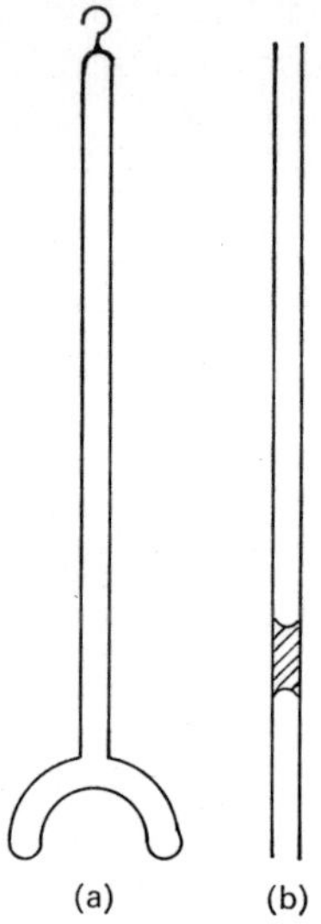

FIG. 5.2. Anti-bumping sticks.

Size of boiler flask. An ether extraction frequently results in a very dilute solution; in many cases no more than 5 per cent of the original bulk of ether solution may remain after the solvent has been distilled off. Consider the situation that could arise, for example, when dealing with 500 ml of an ether solution containing 25 g of solute. To distil off the ether from all of the solution at one time would require a 1-litre flask and there will be excessive drainage losses when removing the residue. To avoid this, the distillation may be done in batches, replenishing the supply of solution to the boiler flask as necessary through a dropping funnel, but the difficulty of using a small boiler flask fed through a dropping funnel is that the rate of boiling, and hence of distillation, is necessarily limited and the process is unduly time consuming. Frequently the best method of handling very dilute solutions is to distil off the bulk of the ether in one batch from a large boiler flask, and, when only a small volume remains, to transfer it and complete the distillation from a smaller flask. The original boiler can be rinsed with ether and the ether washings transferred to the small flask. If this procedure is done carefully the drainage losses are negligible and the overall rate of distillation is considerably increased.

Anti-splash heads. Very rapid distillation that is often possible with volatile materials such as ether, introduces the possibility that the stream of vapour leaving the boiler flask and entering the condenser at high velocity may carry with it some of the solution, entrained as a mist of very fine droplets. If this happens, some of the solute, that should all remain in the boiler flask when solvent stripping has been completed, will be lost by transfer to the ether in the distillate receiver. A similar source of loss is foaming that sometimes occurs especially when solvent extracts of some natural products are boiled.

An anti-splash head will help to eliminate these difficulties, and a typical design is shown in Fig. 5.1. Its main mode of action is to provide an expansion chamber for the vapour, so that its velocity is decreased, and also to provide a baffle surface, at which the vapour stream is made to undergo rapid changes of direction. These two effects delay the emergence of the vapour into the condenser and provide time for suspended liquid droplets to settle out from the vapour and run back into the boiler flask.

Condensers. Considering the condenser as a heat exchanger (see p. 57) the main point to remember is that the rate of heat transfer is proportional to the temperature gradient between the hot incoming vapour and the water-cooling medium, and also to the area of contact between the hot vapour and the cooled surface. When distilling ether the heat-exchange requirements are rather exacting because of the low boiling point of the distillate and a Liebig condenser would have to be inconveniently long to achieve the required rate of cooling. One of the more elaborate multisurface condensers should be used.

Even when a suitable condenser is used the condensed solvent collected in the receiver flask is still apt to evaporate if the ambient temperature is high. To guard against this the receiver should be cooled in iced water or even, in some cases, solid carbon dioxide and acetone (see p. 63). For maximum effectiveness the condenser outlet should lead well down into the receiver flask which should be deeply immersed in the cooling bath. As a further precaution against fire resulting from escape of inflammable vapour the condenser should be sealed with a cork, bung or ground-glass joint into the neck of the receiver flask which should be vented from a side arm by means of a long rubber tube leading down to floor level (Fig. 5.3).

It is frequently necessary to carry out this distillation process on an ether (or other) solution that has been dried. In this case all the glassware used for the distillation, including the condenser and receiver flask should be dry before starting. This is necessary because sometimes, in spite of all precautions to avoid it, some of the boiler content distils too rapidly and froths over into the receiver;

when this happens the only satisfactory treatment is to return all of the material in the receiver back to the boiler flask and repeat the distillation. If any of the apparatus was not dry, it would be necessary to re-dry the distillate before repeating the distillation.

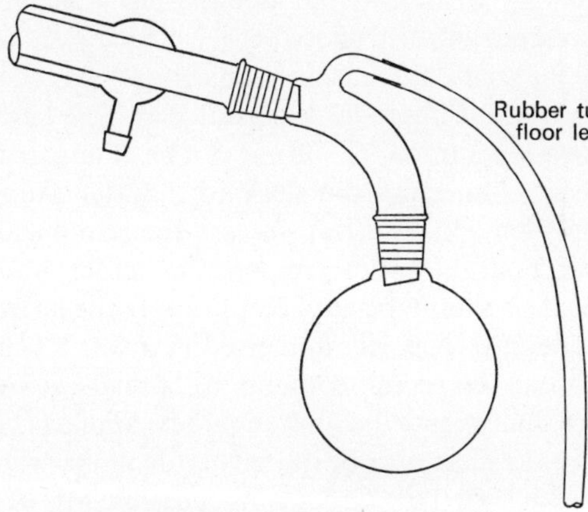

FIG. 5.3. Ether distillation assembly.

Size of receiver. Because there is always the possibility of the contents of the boiler flask being carried over into the receiver, it is good practice to use a receiver flask large enough to hold all of the liquid in the boiler, even though it may be intended only to distil off a small proportion of the total volume. If it becomes necessary to return the distillate back to the boiler flask for redistillation, it should be remembered that the condenser itself will probably still contain some of the solution. Before returning the receiver contents to the boiler flask, normal distillation should be allowed to continue for a short time to provide enough distillate to wash out the solution from the condenser. Then, having returned the distillate to the boiler flask the receiver should be rinsed with some more dry ether and the rinsings returned to the boiler. Finally more anti-bumping stone is added and the distillation resumed.

Removal of Final Traces of Solvent—Stripping

The normal distillation process so far described is effective for removing most of the solvent, but not the last traces from the residue. The presence of this residual ether, which can sometimes be detected by the smell, is frequently objectionable, especially if, for example, the residue has to be re-crystallized. There are various ways in which residual traces of solvent can be eliminated. This is best achieved by connecting the flask to a water pump, via a Buchner flask which acts as a trap in case the pump should fail. Although ether boils at room temperature under water-pump vacuum, gentle heat should be applied to supply the latent heat of evaporation. The flask is partly immersed in a warm water bath, and rotated to encourage the contents to spread as a thin film over the surface. Only gentle heating should be applied (say 40°C), as the residue in the flask may be quite volatile under water-pump pressure at, say, 100°C. To guard against possible loss of material down the pump due to foaming, it is desirable to fit a splash head between the flask and the vacuum connection. This stripping operation is very conveniently done in a rotary evaporator, in which the flask containing the material to be freed from solvent is slowly rotated so as to present a constantly renewed thin film from which solvent is readily disengaged by the combined effects of gentle heating and reduced pressure.

STEAM DISTILLATION

This very useful technique permits the distillation of some high molecular weight compounds from a mixture, at temperatures below their normal boiling points, without the use of vacuum. The detailed principles of steam distillation based on the law of partial pressures and the phase rule are described in most textbooks of physical chemistry. If steam is kept in contact with a water-immiscible organic material, and allowed to equilibrate in the vapour phase, the resulting vapour will be at a temperature just below 100°C, and consist of a mixture of the two components. The proportions of the two components will be in accordance with the following equation:

$$\frac{W_x}{W_w} = \frac{M_x}{18} \times \frac{P_x}{P_w},$$

where W_x and W_w are the amounts by weight of the organic component and water respectively; M_x is the molecular weight of the organic component; and P_x and P_w are the partial vapour pressures of the organic component and water at the temperature at which distillation takes place. It follows that the sum of P_x and P_w is the atmospheric pressure at the time of the distillation.

For very high boiling point compounds, super-heated steam is sometimes used.

Technique

The distillation is usually carried out by injecting steam into the organic material, from a main or a boiler such as a metal can containing water, which is heated directly by a Bunsen burner without a gauze so that maximum rate of steam production is attained. The boiler must be fitted with a safety device, to prevent pressure building up, if the steam outlet should become blocked. This is simply done with a long glass tube (4–5 mm bore) fitted through the bung in the seal of the can, passing down to within about 1 cm of the bottom, and protruding about 50 cm above the neck (Fig. 5.4). If excess pressure builds up in the boiler, the water level will rise in the vent tube and give visible warning that a blockage has occurred. The pressure will be automatically relieved, if necessary, by blowing water out of the vent. This is preferable to an explosion, but there is still the hazard of spraying boiling water. The boiler pressure should therefore be watched and vented, if necessary, by disconnecting the lead from the steam outlet to the distillation flask. The advantage of having the vent

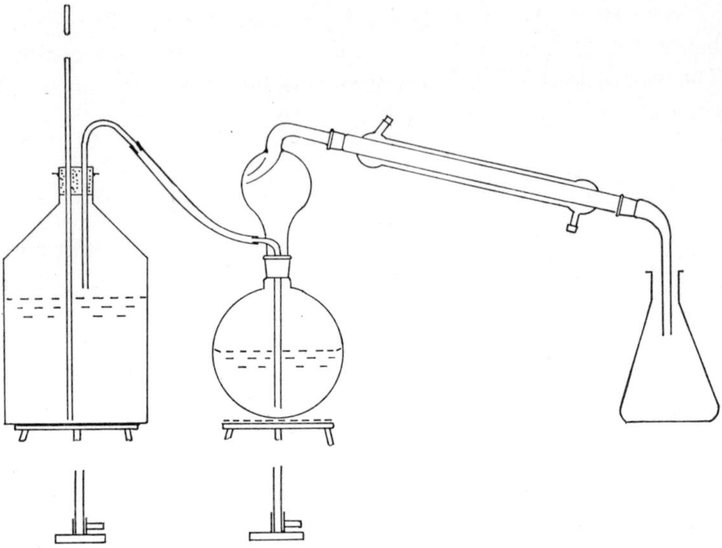

FIG. 5.4. Steam distillation assembly.

reaching only to within 1 cm of the bottom of the can is that, should the water level fall below that, the seal will open and steam will emerge from the vent, providing a spectacular demonstration of the need to refill the can before it boils dry.

Excessive accumulation of water, resulting from the condensation of steam, will occur if steam is injected into a cold organic liquid. The safest way of starting the distillation is to warm the organic material to about 70–80°C on a steam bath (use a thermometer), and then discontinue heating until the steam injection has been started. After a short time, during which some water accumulates in the distillation flask, boiling starts and condensate is obtained. The boiler flask should then be heated at such a rate that the level of liquid in it remains almost constant.

Steam condensing in the feed line between the boiler and the distillation flask may lead to excessive water in the distillation flask. Steam free from water droplets is required, and a trap should be placed in the line just before it enters the distillation flask, especially if being taken from a main on the work bench. The simplest type of trap is made from a separating funnel (Fig. 5.5(a)); water accumulating in the funnel can be run off occasionally by opening the tap. Because it is not self-venting, excessive steam pressure could cause it to burst, and it would be

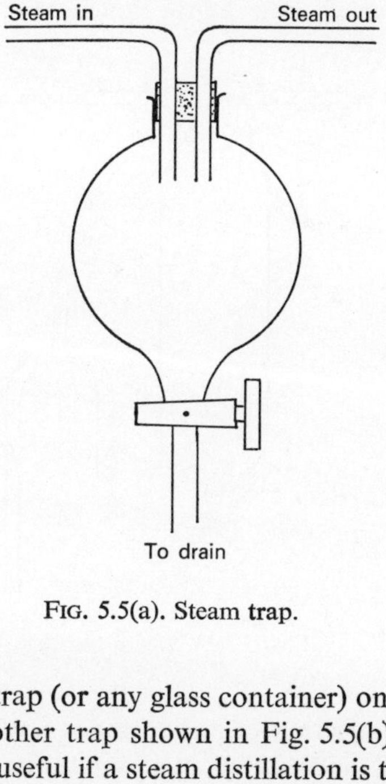

FIG. 5.5(a). Steam trap.

unwise to use this trap (or any glass container) on a high-pressure steam main. The other trap shown in Fig. 5.5(b) is self-venting, and is particularly useful if a steam distillation is to be carried out over a long period of time; but it suffers from the defect that it will blow out its water seal if pressure inside builds up.

Steam distillation requires an equilibrium between the steam and the immiscible organic component, and this can only be achieved by the maximum contact between the two components. This necessitates vigorous agitation, which is best produced by passing a very copious supply of steam into the organic material. The steam inlet tube must reach to the bottom of the distillation flask; and it is usual to fit the flask with an efficient splash head, since liquid droplets can easily be carried over under the vigorous boiling conditions and foaming often occurs. The flask should never be more than one-third full of liquid at the start of a steam distillation, and if an anti-splash head is not available, the boiler flask should be set at an acute angle to the condenser, to minimize risk of liquid splashes entering the condenser.

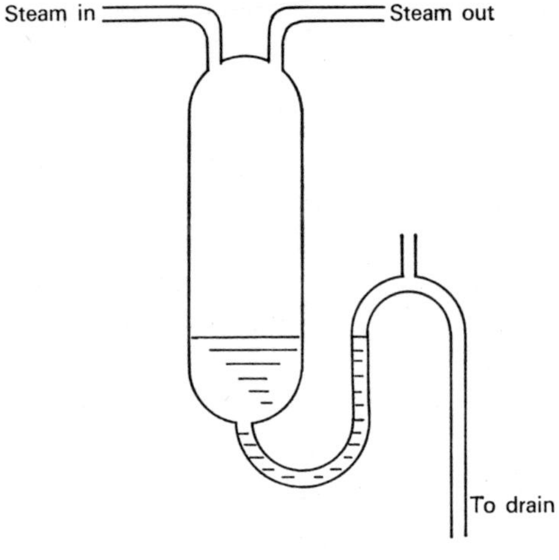

FIG. 5.5(b). Steam trap.

The vigorous injection of steam imposes a heavy load on the condenser, and a multi-surface type should be used if possible. A simple Liebig condenser would either have to be inordinately long to give the desired performance, or it would impose an unnecessary restraint on the rate of distillation.

In the course of the distillation the organic compound will gradually be transferred to the receiver, but the volume of liquid in the distilling flask remains virtually constant, and so gives no guide to the progress of the operation. While the organic compound is present in the distillate, it will appear milky, and will tend to separate into two layers; distillation is therefore continued until the distillate is quite clear, i.e. it consists of water only. It is difficult to see this while the condensate is dripping from the condenser into the main receiver, and about 1 ml should be collected separately in a test tube for inspection. In some distillations the distillate continues to be faintly opalescent for a very long period of time, and it may not be worth spending a lot of extra time to obtain what may be only the last 1 per cent or so of the total organic component.

When bringing a steam distillation to a close, precautions must be taken to prevent the contents of the flask being sucked back into the steam line or boiler as the latter cools. The usual procedure is to remove the heaters and immediately open the connection between the steam boiler and the distillation flask, taking care to avoid scalding.

The distillate obtained from steam distillation will, of course, consist of water and the organic compound, in two separate phases. If some of the organic material remains suspended as fine droplets in the aqueous layer, and clean separation in a funnel is impossible, the techniques of extraction described on p. 121 should be employed.

It is sometimes possible, in small-scale work, simply to add water to the organic material and steam distil the heated mixture, without using a separate steam generator. Care must be taken to provide adequate mixing, and to avoid overheating. The ease with which this technique can be applied very often depends on the density of the organic material. If the organic phase tends to float, steam generation from the aqueous layer below will provide the necessary agitation, and there will be less danger of overheating. If the organic layer is denser than water, this technique can only be applied if stirring is provided; if great care is not exercised there is a distinct possibility of damage to the lower organic layer while heating to induce steam evolution from the water.

FRACTIONAL DISTILLATION AT ATMOSPHERIC PRESSURE

This is a much more exacting technique than distillation for solvent removal, and the following treatment of the subject is not comprehensive: this discussion will be limited to those aspects which are of importance for first-year student operations. Further reading on the physical principles of the process is recommended.

Fractional distillation is a process used for separating the components of a mixture, taking advantage of the fact that their boiling points differ, so that by careful control of the heating conditions, it is possible for the substance with the lowest boiling point to exist in the vapour phase, while the others remain liquid; the vapour can then be led away and condensed. When all of that component has been removed, the heat input is increased until the material with the next higher boiling point can just be converted to the vapour form and that too is collected separately; and so the process continues until all the components of the original mixture have been separated.

In practice, it is usually very difficult to achieve the separation along the clean lines indicated above. For certain mixtures, it is impossible to separate the components, even though their boiling points may differ considerably. These mixtures, which have a definite composition and a constant boiling point, are called azeotropes and are discussed on p. 113.

The principles and practice of a relatively simple fractional distillation can be illustrated by reference to a mixture of acetone (boiling point 56·5°C) and water (boiling point 100°C). These two components do not form an azeotropic mixture and so their separation by fractional distillation is feasible.

As the boiling point of these components differs by about 44°, it would seem that, on heating the mixture to 56·5°, the acetone would start to boil; when all of it had been converted from liquid to vapour and led away to the condenser, pure water would remain behind in the boiler flask. In practice the position is not quite so simple.

If simple distillation apparatus, as for solvent removal is used, the distillation proceeds as a continuous process. The distillate at the start will be reasonably pure acetone, but its composition will gradually change, becoming increasingly rich in water, until towards the end of the distillation almost pure water is collected. Although some pure acetone and some pure water are obtained, much of the distillate would comprise a large volume of "middle fraction", containing both components. To achieve complete separation the middle fraction would have to be redistilled many times.

The lack of clean separation is due to the fact that, in order to obtain proper separation, the liquid and vapour components of the system must at all times be in thermodynamic equilibrium. This state is theoretically impossible to achieve if any distillate is being collected, because removal of even the smallest amount of one of the components of the mixture results in slight disequilibrium. If removal of the distillate is slow it is possible to obtain conditions which are at any point of time not far from equilibrium, so that the separation of most of each of the pure components in a pure state can be achieved. To speed up this equilibration process, a fractionating column is used. Different types of column are described later (see p. 82) but they are all designed to provide a very large ratio of surface-area to volume.

Hot vapour ascending the column is slightly cooled so that the higher boiling component (water) condenses, and runs back towards the boiler, while the vapour continues upwards, becoming correspondingly richer in acetone. As the vapour stream moves up the column, it has to pass through the down-flow of condensed liquid (relatively poor in acetone), and tends to lose water by condensation; at the same time it forms an equilibrium mixture with the liquid running down. The overall effect is the same as if the mixture were re-distilled many times. Equilibration takes an appreciable length of time; moreover, it has already been pointed out that the mere fact of taking off any of one component disturbs the equilibrium. Therefore, the take-off rate must be very slow: theoretically the column can only work at its maximum efficiency when there is no take-off at all. Such operating conditions are of no practical value, but nevertheless fractional distillation cannot be done hurriedly or carelessly.

Boiler Heating

Because fractional distillation is based entirely on establishing equilibrium conditions, it is obviously important to have a steady rate of heat input into the liquid being distilled. Under ideal conditions, this input should only be allowed to change when going from one fraction to another. It is virtually impossible to achieve steady heating conditions by direct application of a Bunsen burner, and one of the more effective methods of heat control described earlier (see p. 59) should be used. An electric heating mantle with a controller permitting adjustment of the rate of heating may be used for high temperatures or for relatively large-scale operations from about 500 ml upwards, but for small-scale distillations a bath containing a suitable hot liquid, and fitted with a thermometer and a stirrer is most satisfactory.

A useful refinement is to introduce a thermometer into the liquid in the boiler, and this is easily done if a multi-neck flask is used. Although the vapour temperature is the most important when carrying out fractional distillation, it is useful to know the temperature of the boiler contents, especially as it provides the first indication of the end of one fraction and the beginning of the next higher one.

Fractionating Columns

A fractionating column must meet two essential requirements: the surface area must be as large as possible to provide the maximum contact between the ascending vapour stream and condensate running down back to the boiler; at the same time, vapour flow from the boiler towards the top of the column must be as free as possible. To some extent these two requirements are incompatible, and the design of a fractionating column is, therefore, a compromise. Two types of column are used for small-scale laboratory purposes—packed and unpacked.

There are many different types of packings for fractionating columns, one of the simplest being Raschig rings (Fig. 5.6(a)),

FIG. 5.6(a). Raschig ring.

which are short lengths of glass or porcelain tube, with the length approximately equal to the diameter. The Fenske packing (Fig. 5.6(b)), consisting of a very large number of single- or multiple-turn helices, usually made of glass, is more efficient. Generally speaking, the smaller the packing material the greater is the efficiency, because of the increased surface area provided, but this advantage is offset by excessive obstruction to the flow of liquid and vapour if the packing is too small.

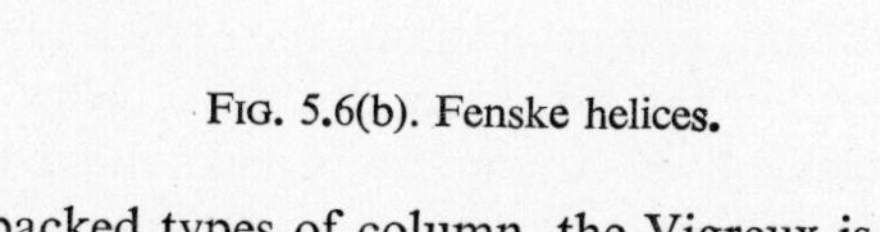

FIG. 5.6(b). Fenske helices.

Of the unpacked types of column, the Vigreux is probably the most useful. The design is based on a cylindrical glass tube into which are made a series of indentations, providing a large surface area to be wetted by the vapour and condensate flow (Fig. 5.7).

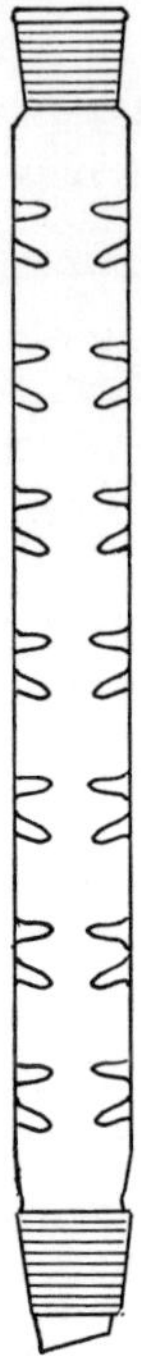

FIG. 5.7. Vigreux fractionating column.

Principles of Column Operation

The efficiency of a column is assessed by its effectiveness in the separation of multi-component mixtures. The quantitative bases of designing columns and determining their efficiency need not be considered here, but the principles described below will make it possible to use any given column to its maximum efficiency.

A fractionating column must be set up so that it is truly vertical when viewed from all directions, so as to secure the maximum possible contact between vapour and liquid flow in the column. Fig. 5.8 shows in an exaggerated way the effect of not having the column vertical; vapour may travel up the higher side of the column and fail to come into contact with the downflow of liquid, along the lower side of the tube.

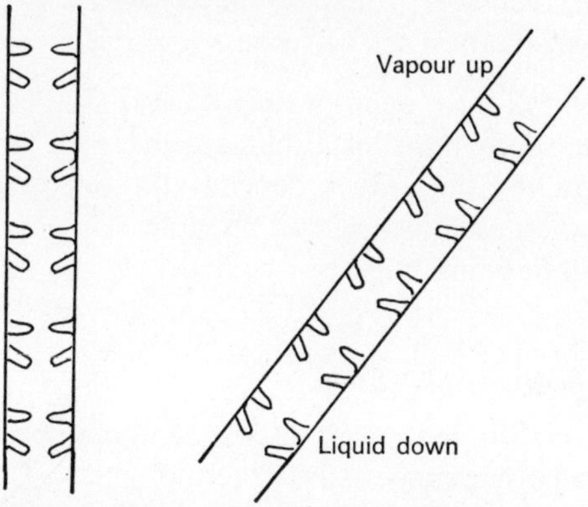

FIG. 5.8. Effect of tilt on fractionating column.

Boil-up Rate

The essential function of a fractionating column is to provide a scrubbing action by the down-flow of condensate upon the vapour moving up. It is obvious, therefore, that there must always be completely free passage for these two streams. In any column, but especially in packed ones, the rate of boiling is critically important; as this is increased, the quantity and velocity of vapour moving up and of liquid running down both become greater, until a point is reached when the vapour velocity is so high that it totally impedes the downward flow of liquid into the boiler flask. This produces a condition known as flooding, and is apparent when slugs of liquid are seen completely to block parts of the cross-section of the column. If this condition is allowed to develop, the increasing pressure of the liquid in the column will cause further difficulties, such as blowing out of stoppers in the boiler flask. Flooding of the fractionating column must, therefore. always be avoided; once it has started, the column ceases to function as a means for establishing equilibrium between vapour and liquid. In normal use it is customary to have the rate of boiling in the flask as high as is possible, provided column flooding does not occur.

To sum up, the main principles underlying the use of a column in fractional distillation are as follows:

(1) The column is an equilibrating device.
(2) Equilibration is not instantaneous.
(3) The equilibrating effect depends on the rate of boiling, which should be as rapid as possible.
(4) Column flooding must be avoided.

Vapour Temperature Indication

For all normal student work, vapour temperatures are indicated by conventional mercury-in-glass thermometers. The purpose of the thermometer is to indicate the temperature of the vapour as near as possible to the point of take-off from the column. It must, therefore, be placed carefully in the column, in such a way that the bulb, which is the only responsive part of the thermometer, is fully exposed to the vapour stream. This is illustrated in Fig. 5.9,

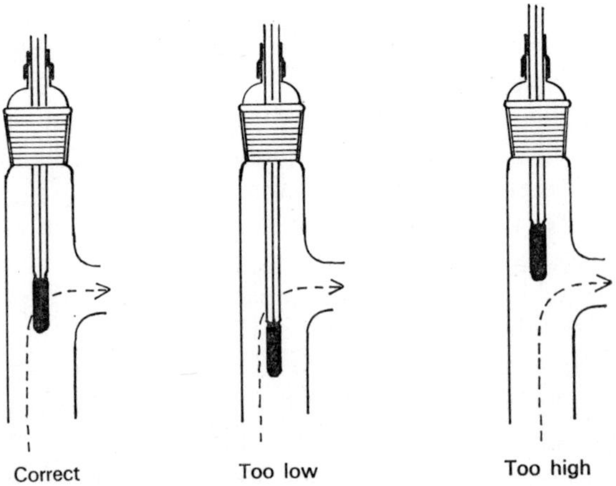

FIG. 5.9. Position of thermometer for distillation apparatus.

which shows, in an exaggerated way, how the thermometer may be wrongly located in the fractionating column.

Choice of Thermometer

The accuracy of the calibration of a commercially produced thermometer can be assumed for elementary work, but some safeguards are necessary if greater precision is required. Thermometers are calibrated in various ways. In some cases, the calibration is based on "total immersion": if such a thermometer is used with only the bulb and a short length of the column exposed to the vapour most of the mercury column is at a lower temperature than that used for calibration, and the reading will be slightly low. A "stem correction" can be made, but it is more convenient to use a thermometer calibrated for an immersion of about 5 cm or as is appropriate to the dimensions of the apparatus being used. Most thermometers are marked to indicate the length of stem that should be exposed to the temperature being measured, e.g. "tot. imm" or "5 cm imm".

The bulb of the thermometer should be as small as possible to minimize the amount of heat it must itself absorb in responding to a rise in temperature. This is particularly important in a small-scale fractional distillation, because this heat absorption results in a slight change of thermal equilibrium in the system being measured. This introduces the need for a compromise; in order to obtain the maximum sensitivity of the thermometer system—that is, the maximum speed of response to the smallest temperature change—the smallest possible quantity of mercury should be present in the bulb. On the other hand, such a small amount of mercury will necessitate very close spacing of the calibration marks on the stem, which are then difficult to read.

For certain high precision work it may be necessary to carry out a special calibration, or to rely upon a thermometer supplied with a calibration certificate, which in the U.K. is issued by the National Physical Laboratory.

Sets of very high sensitivity thermometers (Anschutz) are

available, each indicating a relatively small temperature range of perhaps 50°C; these are expensive and generally reserved for special work.

Control of Reflux Ratio

This is very important. The reflux ratio is defined as the ratio of the volume of liquid running back (refluxing) to the boiler, to the volume being removed as a distillate. In the majority of simple operations, a reflux ratio of at least 10:1 should be maintained. There is no theoretical objection to a very high reflux ratio: it has already been shown that, in theory, the maximum possible efficiency of a fractionating column is only obtained when equilibrium has been established and is not disturbed by removing any distillate. Under these conditions, the reflux ratio is infinity, and although this will give maximum possible efficiency, it is obviously of no practical value.

The measurement of reflux ratio is simply done by counting the number of drops returning from the bottom of the column to the boiler and comparing that with the number collected in the same time as distillate from the end of the condenser. In many cases, the problem of controlling reflux ratio solves itself. With a Vigreux column, for example, the normal atmospheric cooling of the column gives a sufficient amount of condensation to provide the necessary reflux, and adequate control of the whole of the fractionation operation can be achieved by adjusting the input of heat to the boiler flask. In distillations involving high boiling fractions (100°C or higher) too high a reflux ratio is obtained as a result of excessive cooling of the outward vapour stream in the column, and if this trouble arises, the only cure is to reduce or eliminate heat losses from the column. In some cases, external heating must be applied by an electric heater, but lagging will usually be sufficient; for a small column, this is most conveniently done by winding with asbestos cord (about $\frac{1}{4}$ in. diameter). It is useful to keep these columns permanently lagged and the asbestos should then be further protected with a winding of linen tape. As a temporary expedient a column may be lagged with cotton wool, which is quite effective, but inflammable, and great care must be taken, especially if a bare flame is being used as part of the heating system. A minor drawback of a lagged column is that unless small inspection windows are cut, it is no longer possible to see what is happening inside it, and incipient flooding is not detectable.

For small equipment, very efficient heat insulation which does not interfere with visibility is provided by a vacuum jacket round

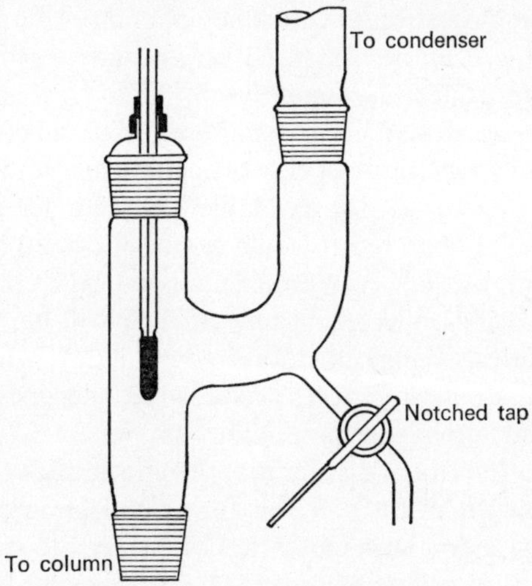

FIG. 5.10. Fractional distillation head.

the column; such columns are commercially available. It is important to remember that reasonable efficiency is only obtained when the internal pressure in the vapour jacket is not more than 0·01 mm of mercury. For the maximum possible heat insulation effect, the vacuum jacket should be internally silvered, leaving windows in the silver coating for inspection of the column.

Control of reflux becomes more important when using very high efficiency columns where ratios of 20:1, or even more, are necessary. This can be achieved, together with a high boil-up rate with the aid of a reflux splitter which returns a large proportion of the condensed vapour from the condenser down the column to the boiler flask, while the smaller portion is continuously taken off; the proportions can be varied (Fig. 5.10). The student is unlikely to require this type of equipment in elementary work.

Fraction Collection

In most cases, a steady boiling point in fractional distillation is regarded as an indication of constancy of composition, but refractive index, which can be measured very rapidly, provides a much more sensitive guide. The method of taking measurements of refractive index is described in detail on p. 176, and corrections for the temperature variations of this property are also discussed.

If a refractometer is not available, the changes in refractive index can still be observed to some extent and used as a guide to change in composition. Any major change in the composition of the distillate being collected in one receiver will be accompanied by corresponding change in the refractive index, which becomes apparent as a streaking effect, produced as successive drops run into the liquid already accumulated; this effect is quite readily observed. In a fractional distillation, the appearance of this streaking in the contents of the receiver during collection of any but an intermediate fraction is an indication of the need to start collecting a new fraction, even if no change in temperature has been noted.

When one particular fraction has come to an end, it is necessary to increase the heat input to the boiler in order to vaporize the next most volatile component. Theoretically, there should be a complete interruption in the distillation between the end of one fraction and the beginning of another, but in practice this condition is never achieved, even with a very high efficiency column, and it is impossible to avoid obtaining some small amount of distillate, whose boiling point lies between that of the two fractions in question. This intermediate fraction must always be collected separately; it is only permissible to start collecting a new main fraction when the boiling point and the refractive index have settled down to a steady value. It must also be remembered that, although a new, pure fraction may be indicated by the vapour temperature, some of the preceding (intermediate) fraction will still remain as a liquid film in the condenser and on the other surfaces between the thermometer and the receiver. Time must therefore be allowed for these to be flushed out by the flow of the next fraction before the main collection is started. It follows from this, therefore, that the distillation apparatus should have the smallest possible area of glass requiring this washing-out treatment, and all condensers and receiver adaptors should be as small as is practicable. In most simple fractional distillations at atmospheric pressure, distillate can be collected directly from the end of the condenser by placing a receiver in a suitable position, and it is not normally necessary or desirable to use a "receiver adaptor", which serves no useful purpose and merely provides another glass surface to be wetted by, and to waste, the distillate. A detailed discussion of the techniques involved in collecting the separate fractions is given on page 95.

As a further safeguard against contamination of fractions, the condenser should be set at a very steep angle from the horizontal, so as to promote maximum possible drainage from it into the receivers. It has already been emphasized that the column must always be vertical, and therefore the angle of the condenser is mainly determined by the design and construction of the vapour take-off fitting attached to the column, but if a bung or cork is used to connect the two, the way it is fixed and bored can provide for a certain amount of adjustment.

Practical Details

The following notes refer specifically to the fractional distillation of a mixture of approximately equal volumes of acetone and water, but they can be adapted to any other particular case; they should be regarded as a guide only, and not a rigid drill. At all times the procedures must be carried out as an application of the physico-chemical principles already explained.

Assembly of apparatus. Special attention should be paid to the following points:

(1) The empty boiler flask is weighed, so that the weight of any undistilled residue may be determined. It should be half to three-quarters full at the start of distillation.

(2) The receiver flasks are cleaned, dried and weighed before the distillation is commenced. If they are to be used to collect volatile fractions (b.p. up to about 60°C) they should be immersed up to the neck in iced water during the collection of the distillate.

(3) An oil bath is best for heating the boiler flask which should be immersed almost up to the neck. The bath must be provided with a thermometer, and provision made for stirring the oil.

(4) Two or three (not more) pieces of porous pot or anti-bumping granules are introduced into the boiler flask.

(5) The column is carefully adjusted so that it is perpendicular. This may be done by visual inspection, viewing along two directions at right angles; a plumb-line or spirit level may be used.

(6) The condenser must be fitted so that it is at the maximum angle to the horizontal, for reasons already given on p. 91.

(7) The water is connected so that it enters the condenser at the lower end (away from the vapour entry).

(8) The thermometer is fitted into the head of the fractionating column, so that the bulb will be just swept by the vapour stream as it is about to enter the condenser. The bulb of the thermometer should be as small as possible. Before starting the distillation, the apparatus is examined to see that temperature readings will not be obscured by the fitting securing the thermometer to the top of the column. If a cork is used for this fitting, difficulty can sometimes be avoided by cutting away surplus material. If this is not enough to eliminate the difficulty, the only satisfactory cure is to change to a more suitable thermometer.

Initial heating. The temperature of the boiler contents is raised until boiling commences, and vapour starts rising in the column. If an oil bath is used, the oil is stirred and its temperature noted; the latter will tend to be 20–30° higher than that of the boiler flask contents. If an electric heating mantle is used, full power is applied initially, and this is reduced by about 90 per cent at the first sign of boiling in the flask. In all cases, the heat input must be carefully controlled once boiling has started, so that vapour in the column, which appears as a striated band, moves only slowly up the column. Allowance must be made for the heat capacity of the heater system, especially with an oil bath, which responds only slowly to a change of power input.

At all times the column must be examined to be sure that flooding does not occur; under proper working conditions there should be no evidence of large amounts of liquid in the column, apart from the presence of a film moving over the surface of the column and its packing. Flooding first becomes apparent at the base of the column as an accumulation of liquid, filling the whole of the cross-section, and disturbed by the vapour rising from the boiler flask. At any stage in a fractional distillation, at the first sign of flooding, heating must be discontinued, and not resumed until the column has drained completely. It is not sufficient merely to reduce the power supply to the heater, as the response will be too slow. The apparatus should be assembled in the first place to permit the lowering of the bath or mantle away from the boiler flask: this is much more satisfactory than trying to move the distillation apparatus.

Having established the initial rate of boiling at a satisfactory level, heat input may be increased cautiously, subject always to avoiding column flooding, until the vapour stream just surrounds the thermometer bulb, so that a drop of liquid condenses on it. This is a useful guide to correct distillation rate; if the heating is properly adjusted, a drop of liquid continues to hang from the bulb. If the vapour flow is excessive, the drop will be carried away; if insufficient, the evolution of distillate will be erratic. The temperature rises rapidly as the vapour surrounds the thermometer bulb, and should be allowed to settle to a steady reading before collection of the first main fraction is commenced.

Subsequent procedure depends on whether or not the column is fitted with a fractionating head. If it is, heating is adjusted to give the maximum flow of liquid through the column, just short of flooding. Initially the fractionating head is adjusted for zero take-off, until equilibrium is established, as indicated by the steady reading of the thermometer. The take-off rate is then changed to permit very slow removal of distillate, so that the reflux ratio is at least 20:1. In any case, the rate of take-off should not be more than about one drop in 5 seconds.

If a fractionating head is not fitted to the column, the same rate of distillation should be aimed at, but here the boil-up rate is much more critical, and under satisfactory operating conditions, the column will probably be operating well below its flood point.

Slow steady collection of this first distillate fraction, known as "forerun", should be continued into the first receiver until the whole of the glass surface has been wetted with distillate whose boiling point is at a steady value. At this point, the receiver is replaced with another clean, dry one which has been weighed.

Consider the position that now develops. The heat input has been adjusted so that it is just sufficient to permit slow distillation of the material boiling at 56·5°C. When all of this has left the boiler flask, the same heat input will be insufficient to maintain the rate of boiling for the remainder of the contents, because these boil at a higher temperature. Distillation will therefore tend to stop, and because the vapour supply through the column is not being maintained, the thermometer is no longer exposed to a steady vapour stream, and the indicated temperature falls. At this stage, the heat input is increased to promote boiling of the next higher fraction (water in this particular example).

Here a complication can arise. The vapour is once again made to rise from the boiler, and an intermediate condition is reached; although water vapour is entering the base of the column, it pushes up ahead of it the last traces of acetone, which were already in the column when distillation began to slow down. Thus when the heating is increased sufficiently to bring up the next higher fraction, and when the distillation is resumed, the material collected is the end of the acetone fraction. The boiling point will therefore settle momentarily at about 56°C again. Quite soon, the water vapour that entered the base of the column will have climbed up to the thermometer level, and the vapour at that point will be a mixture of acetone and water: the temperature will climb beyond 56°C, and the distillate coming over is neither pure acetone nor pure water. This intermediate fraction is collected in a separate flask, at an even slower rate than before (say one drop in 10 seconds) while the heat input to the boiler is very slowly increased. The aim now is to establish a new equilibrium in the column, while collecting only the minimum quantity of distillate. This process is continued until the distillation temperature once again settles at a steady level—100°C in this case.

Consider the situation at the moment when the thermometer first indicates 100°C. Two points must be allowed for. First, distillation must be allowed to continue until it is certain that a steady temperature has been reached. Secondly, when the first drop of pure water enters the condenser, the rest of the glass surfaces between the column and the receiver will still be wet with the intermediate fraction. Collection of the new fraction is only permissible when the water has been distilling for long enough to flush out all the surfaces between the thermometer and receiver.

Change of fraction. It is sometimes difficult to decide exactly when to change the receiver, to collect the next higher fraction. This decision involves a compromise between the highest purity of the fraction being collected and the maximum yield. If the purity of the first main fraction is very important and a small loss of yield can be tolerated, the receiver containing this distillate should be changed when the boil-up rate decreases, and before the heat input is increased.

Although temperature is the main guide to the start of the new fraction, even a small change in composition will be accompanied by a change in refractive index and density, which will result in a characteristic striation effect as a distilled droplet enters the receiver. Therefore, collection of a main fraction should always be discontinued, either on the evidence of temperature, or when striation of the distillate is first observed. While collecting the intermediate fraction, only temperature is significant, since by the nature of the intermediate fraction, which is rapidly changing in composition, striation is to be expected.

To illustrate the procedure described in this section, the results of two distillations, carried out with different degrees of skill, will now be analysed. The example chosen is still the acetone–water mixture.

Fraction no.	A Good distillation			B Bad distillation		
	b.p. (°C)	Vol. (ml)	n_D^{20}	b.p. (°C)	Vol. (ml)	n_D^{20}
1	Up to 56·5	0·5	1·3571	Up to 56·5	3·0	1·3540
2	56·5–57·0	49·0	1·3580	57–58	41·0	1·3537
3	57·0–99·8	1·0	1·3417	58–100	13·0	1·3422
4	99·8–100	48·0	1·3333	100	32·5	1·3330
Residue		3·0	1·3338		10·0	1·3332
Total		101·5			99·5	

Fraction 1, the "forerun", leading up to the first main fraction (acetone) should be as small as possible. In A the quantity collected is reasonable considering the time required for a distillation to settle down, and to ensure that collection of acetone is not started prematurely. In B the excessive forerun could be due to too high a rate of heating. The refractive indices for these fractions are revealing. In A the value is only a little lower than for pure acetone. This could be accounted for by contamination with a trace of water, due to atmospheric deposition on the surface of the distillation apparatus: in B the forerun is relatively rich in water, and hence the refractive index is rather low.

Fraction 2 in each distillation represents the main collection of acetone. In A recovery is excellent, the boiling range very close, and the refractive index in good agreement with the literature figure. In B the yield is poor, and the low refractive index suggests contamination with water. The boiling range is quite reasonable, even though it is not quite as good as in A. Most of these criticisms are attributable to an excessive rate of distillation.

Fraction 3 is the intermediate fraction. In A this was collected very slowly, giving the column time to equilibrate, and the amount collected is satisfactorily small. In B the intermediate fraction was distilled much too rapidly, and therefore the amount collected is excessive.

Fraction 4 should be pure water. In A the percentage recovery is excellent; the poor recovery in B is, once again, a consequence of excessive haste in distillation. Some of the water came over in Fraction 3, and some remains in the undistilled residue. Even in the good distillation it was inevitable that some water would be left in the flask, due to drainage from the fractionating column as it cooled.

The total volume accounted for in distillation A may seem surprising: 100 ml of mixture has been distilled, while the total volume accounted for is 101·5 ml. This is explained by the physical principle that when two miscible liquids are mixed, the total volume is not necessarily the sum of the two component volumes; in many cases it is less. In the distillation under discussion this position was reversed. The mixture, where the contraction had already taken place, was "opened up", and hence the sum of the volumes of the components is greater than the volume of the starting material.

Chasers. The residue of the higher fraction which remains in the boiler flask at the end of a distillation can present a problem. It arises because the boiler flask becomes dry and distillation ceases, but on cooling the apparatus, liquid remaining in the column drains back into the flask. This can account for a considerable amount of material that should be part of the final fraction. This problem can be overcome in many cases by using a "chaser"—a deliberately added higher boiling fraction of a liquid chosen because its boiling point is considerably higher than that of the final fraction. It must be chemically non-reactive with the other components, and it must not form azeotropes with them. It is added at the beginning of the distillation, so that at the end it displaces all of the preceding fraction in the column. When using a chaser, therefore, distillation is continued until just short of the point where this high boiling material is about to reach the top of the column.

DISTILLATION AT REDUCED PRESSURE

Where the boiling point of an organic material at atmospheric pressure is too high to allow distillation without decomposition the operating pressure is reduced to lower the boiling point. The principles of distillation are not significantly different at lower pressures, but the use of reduced pressure introduces the need for many special techniques.

One of the problems concerns pressure distribution throughout the system. In any distillation, the pressure should be the same throughout the whole apparatus, and should not change during the distillation: only in this way can boiling point temperatures of any significance be obtained. It is impossible to achieve truly equal pressure distribution during the actual distillation, because at one end of the system the boiler is constantly producing vapour (and hence increasing pressure) which is removed by the pump at the other. Therefore, a pressure gradient exists between the boiler and the receiver.

There are two basic requirements for any distillation under reduced pressure: a pump capable of producing and sustaining the required vacuum, and leak-proof distillation equipment that will give the minimum pressure gradient between the boiler and the receiver.

Vacuum Pumps

Three types of pump were mentioned on p. 66:

Water pump: 11–25 mm of Hg.
Mechanical oil pump: 10–0·01 mm of Hg.
Vapour diffusion pump: about 10^{-3} mm of Hg or lower.

The best vacuum a pump can achieve depends on the principle of its operation. Its pumping speed, i.e. the rate at which it removes air or other vapour from the system and reaches that ultimate vacuum, is, however, mainly a function of its size, and larger pumps have higher pumping speeds.

Water pumps. This type of pump cannot reduce the pressure to below the vapour pressure of water at the ambient temperature. The best vacuum theoretically attainable is about 10 mm Hg: in practice, a water pump will give about 11 mm Hg in winter, and up to about 15 mm at summer temperatures. The pumping speed depends on the dimensions of the pump, and on the pressure and flow rate of the water. Difficulty in using water pumps often results from the fact that they are fed from a header or storage tank which may be no more than about 15 feet above the laboratory bench: the resultant water pressure will certainly be insufficient for satisfactory operation of the pump, not only because the rate of evacuation will be low (which could probably be tolerated) but also because any slight variation in pressure arising from the opening of another tap on the same line will cause failure of the pump, and water will be sucked back into the system being evacuated. A similar trouble is caused by turning off the water supply to the pump while it is still connected to the evacuated system. It must be made a habit to open the vacuum connection *before* the water is turned off; this must be done even if a non-return valve is placed between the pump and the system being evacuated, because these devices are not always dependable. Because of the constant possibility that the water may suck back, a trap should be placed between the pump and the evacuated system. A glass Buchner flask is quite suitable and the connection to the pump should almost reach the bottom of the flask, so that any water entering it may be sucked out again by the pump itself. It is not safe to use an ordinary flat-bottomed flask or a bottle as a trap: unless specifically designed for use under vacuum, such glass vessels are likely to implode when evacuated.

The water pump is particularly suitable for distillations of volatile solvents, because the pumping fluid (water) is continuously replaced, and is not liable to the permanent contamination that oil pumps may suffer. Glass pumps are resistant to corrosion under virtually all conditions, but they are fragile and their performance can vary widely, because of difficulties of manufacture to the close dimensional tolerances required. A glass pump should therefore be checked before use, by connecting it directly to a manometer to see that it is capable of achieving a pressure of 15 mm of Hg (or lower) in a reasonable time. Metal and plastic pumps are more reliable, but can sometimes introduce corrosion problems. Water pumps must always be used with the water supply full on: to control the vacuum, a leak controlled by a rubber tube and screw clip may be introduced into the line connecting to the apparatus. It is not satisfactory to try to achieve this control by adjusting the water supply; suck-back will always certainly occur.

High vacuum pumps. Only the rotary oil pump will be discussed here, as students of elementary work are unlikely to require vapour diffusion pumps.

In any pumping system the best vacuum that can be produced is limited ultimately by the vapour pressure of the pumping fluid, so the efficiency of a rotary oil pump depends upon the use of an oil whose vapour pressure is low at the working temperature. Oils are easily able to dissolve other organic vapours; if this occurs, their effective vapour pressure increases considerably, and there is a corresponding decrease in their effectiveness as pumping fluids. It is therefore essential to take special precautions to prevent the entry of condensable vapour in the pump. This is done with traps, cooled with solid carbon dioxide in alcohol (or acetone) or with liquid air (Fig. 5.11(a) and (b)).

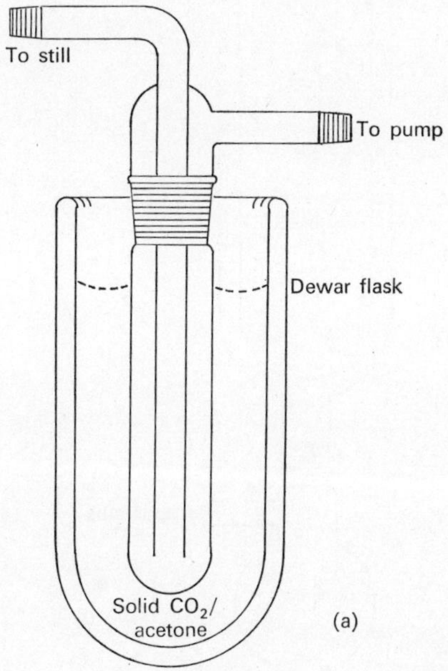

FIG. 5.11(a). Trap for high vacuum distillation.

These traps are designed to remove vapours by condensation at very low temperatures while offering the lowest possible resistance to gas flow. When a rotary oil pump is switched off, air must be admitted to the connection between the main vacuum line and the pump; if the pump is left connected to an evacuated system, the gradual leakage of air into the pump will cause the oil to be pushed up into the vacuum lines, and possibly into the traps and distillation equipment. Another precaution to be taken at the end of a distillation, is to close the connection between the pump and the traps until the latter have been thoroughly cleaned, so as to avoid contamination of the oil by volatile materials diffusing from them as they warm up.

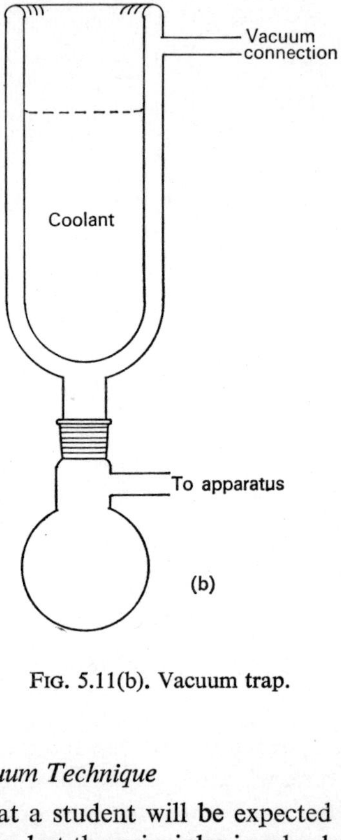

FIG. 5.11(b). Vacuum trap.

General High Vacuum Technique

It is unlikely that a student will be expected to set up a high vacuum installation, but the principles involved should be understood so that the equipment can be used intelligently.

The use of a satisfactory pump does not, in itself, guarantee that the required conditions of pressure will be achieved in a distillation. The "back pressure" of a vacuum system, which is roughly comparable in effect to the viscosity of a liquid system, is illustrated diagrammatically in Fig. 5.12. Vessel A has a small inlet to the atmosphere C, and is evacuated by a pump B. Connection between A and B is provided by a pipe X. Manometers P and Q are fitted. Consider the following cases, in all of which it is assumed that B is pumping at the maximum rate.

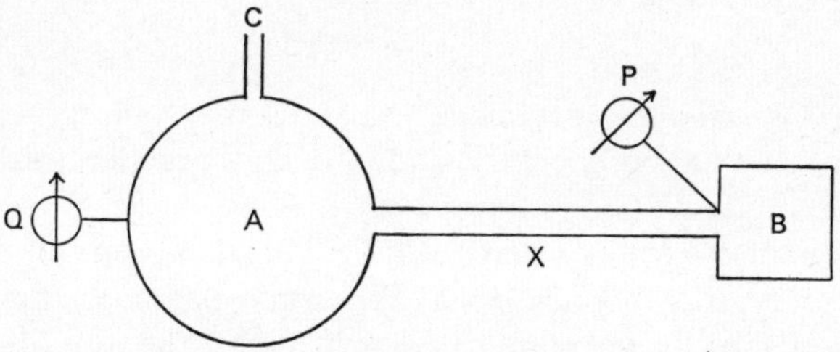

FIG. 5.12. Vacuum pumping system.

Case 1. X completely closed; C open to the atmosphere. Obviously the pressure in A will be atmospheric.

Case 2. X is open, and has a very wide bore. Here, A will quickly become evacuated; manometers P and Q will give virtually identical readings which will depend on the pumping speed and ultimate vacuum that B can achieve, and on the size of leak C.

Case 3. X is open, but consists of a long small-bore tube; or alternatively, X has a wide bore but is packed tightly with a porous material such as sand. The manometer P will show a much higher pressure than Q, because X offers very severe impedance to flow of vapour from A to B; the system is now virtually intermediate between Case 1 and Case 2.

From this illustration it follows that a high vacuum assembly should be made up of wide-bore (1 cm or more) glass tubing, preferably using all-glass (or metal-to-glass) joints. If rubber tubing must be used, it should have thick walls and a wide bore and be completely free from cracks in the walls.

Measurement of vacuum. Vacuum gauges fall into three broad categories, corresponding to the working pressures given by water, rotary oil and diffusion pumps.

Mercury manometers measure pressure directly in terms of the height of a column of mercury, and in general practice are suitable for pressure down to 10 mm of Hg.

The simplest type is a vertical tube with its lower end dipping into a reservoir of mercury open to the atmosphere; the upper end is connected to the vacuum line. The length of the tube above the level of mercury in the reservoir is at least 80 cm. The difference between the height of the mercury column above the level in the reservoir, and the reading of an ordinary barometer, gives the pressure of the vacuum system. This simple manometer is fragile and inconvenient because of its large size and the scale must always be adjusted so that zero coincides with the level in the the reservoir. The closed end manometer is more compact, and gives a direct reading independent of atmospheric pressure (Fig. 5.13). Provided there is only mercury vapour in B, the height of the mercury column will balance the pressure at A. The possibility of air entering the closed space of the manometer above the mercury should always be considered. If this is suspected, it can be checked by applying the gauge to a water pump; if it indicates a pressure substantially lower than 12 mm of Hg, it is certain that the manometer has air, water vapour or some other gas trapped in it, and the instrument must be cleaned and refilled. Once this type of gauge is under a vacuum and showing a pressure of, say, 15 mm, great care must be exercised in opening it to atmospheric pressure; a sudden increase in pressure will cause the mercury to slam up against the closed end of the tube and probably

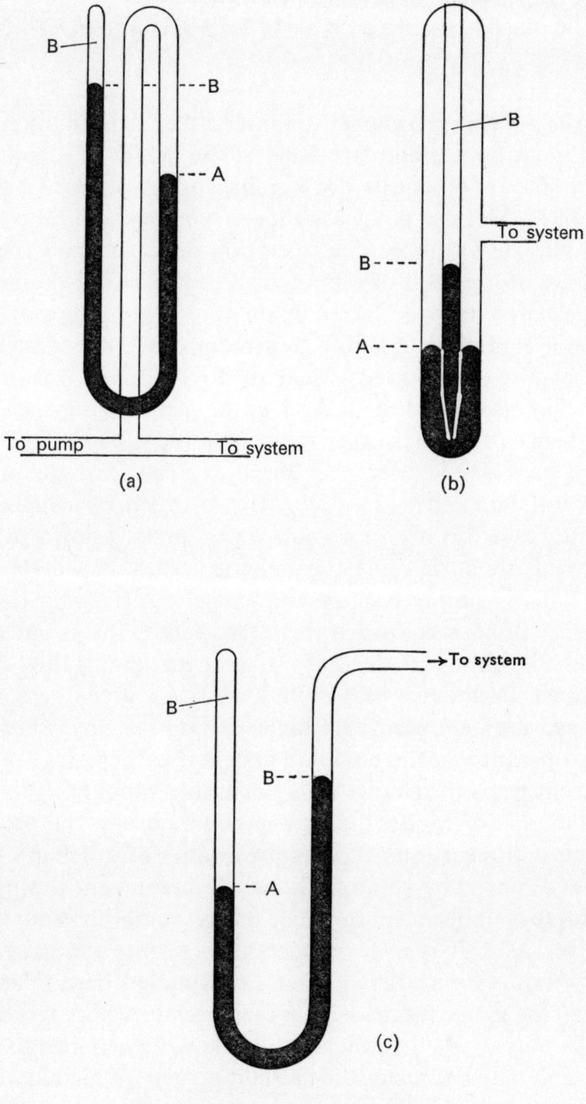

FIG. 5.13. Vacuum manometers.

break it. To protect the gauge from suck-back of liquid, a trap should be placed on each side of it in the vacuum line.

McLeod gauges. The gauges so far described depend upon direct measurement on a millimetre scale of the height of a column of mercury. This is obviously not feasible for pressures of less than 1 mm of Hg. McLeod gauges are therefore used with rotary oil or vapour-diffusion pumps. The operation depends upon the compressibility of a pocket of a known volume of air at the pressure to be measured, trapped by a column of mercury. The gauge does not give a continuous reading, but requires to be operated each time a reading is required; when set to produce a reading, the gauge will not respond to changes of the pressure it is measuring. The presence of any gas other than air in the gauge will give false readings because the original calibration is based on the presence only of the latter. This is easily demonstrated by connecting a McLeod gauge directly to a good water pump, known to give a vacuum of about 11 mm; the reading indicated will be about 5 mm. This is simply because the gauge is measuring the compressibility of air saturated with water vapour; this is much more compressible than the dry gas, so that an unduly low reading is obtained. In practice, this is not a serious source of error, since the gauges are only used with high vacuum systems, where satisfactory operation of the pumping system itself depends on the use of efficient traps to remove all condensable vapours.

In one type of McLeod gauge in common use, the process of trapping and measuring the compressibility of a definite volume of air is achieved by rotating the whole instrument through 90°: this rotatory movement must be made smoothly and without excessive haste. It is most important to restore the gauge to its resting position immediately after a reading has been taken. Only then can the gauge react to changes in pressure, and, moreover, if air is admitted to the system while the gauge is in the "read" position, it may be damaged by a sudden surge of mercury into the calibrated limb.

Electronic devices. Pirani gauges and other electronic devices are used for measuring very low pressures, say 10^{-2} down to 10^{-8} mm of Hg; these are not likely to be encountered except in advanced work.

Prevention of bumping. Under atmospheric pressure, smooth boiling can sometimes be attained without any anti-bumping device, but this is scarcely ever possible when the pressure is reduced. One reason for this is that under the latter conditions the pressure due to the head of liquid in the flask on the vapour produced at the bottom by localized heating, is large in relation to the total pressure in the container: the result is that the vapour tends to accumulate in a closed pocket of liquid on the bottom of the flask and instead of being constantly liberated, it is apt to discharge irregularly, with almost explosive violence.

Porous pot does not prevent bumping at reduced pressure. A more effective device is a long wooden splint or boiling stick (Fig. 5.2). Another method, not highly satisfactory, is to pack the boiler very loosely with glass wool, filling the flask up to, but not including, the neck. This device is only suitable for small-scale operations (up to perhaps 100 cc); it depends on the capillary rise along the fibres of the glass wool above the main liquid surface, to give a large evaporative area. The application of vacuum may cause the glass wool to collapse below the liquid level, and hence become ineffective; this can be avoided by adding more wool. The main objection to this method is that the area of the glass surface to which the liquid is exposed is very large, so that losses resulting from the wetting are excessive. Some organic materials are adversely affected by the traces of alkali present in the wool, which is usually made of soft glass.

If it is not essential to work with an entirely closed system, the most reliable method of promoting smooth boiling in an evacuated system is to use a very fine aspiration tube drawn out from thick-walled capillary tubing as described on p. 15. It is fitted into the flask so that the tip just clears the bottom; this should be done before the liquid is added to the flask, to minimize risk of breaking the capillary. The upper end of the bubbler is left open. When the pressure in the system is reduced, air is drawn through the capillary, and discharges as a stream of very fine bubbles, agitating the boiler contents, and continuously breaking the surface. Unfortunately, this excellent device is limited to distillations where the passage of a small amount of air or other gas through the

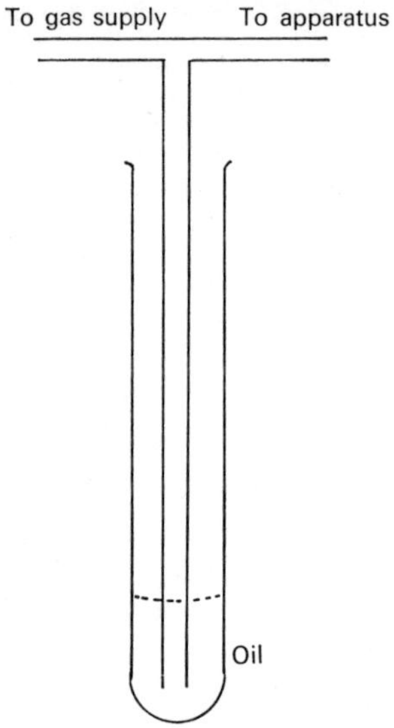

FIG. 5.14. Pressure limiting device.

system is permissible. If the oxygen in the air is undesirable, the external end of the capillary is connected to a source of nitrogen or other inert gas. This must not be done by direct connection to a cylinder or gas generator because the resistance to gas flow offered by the capillary will cause serious pressure build-up. A satisfactory device is shown in Fig. 5.14.

The capillary below the liquid surface cannot be too finely drawn out, but in order to avoid breakage, the tip should be as short as possible (2–3 cm); it is best prepared by first drawing out to a coarse capillary and then again heating and pulling out to a very fine diameter. The leakage of air or other gas into the system through the capillary increases the demands made upon the pumping system. Even a water pump should have a sufficient capacity for use with this device if it is properly made, and there should be no significant deterioration of vacuum. One of the difficulties encountered in use is that when air is readmitted to the

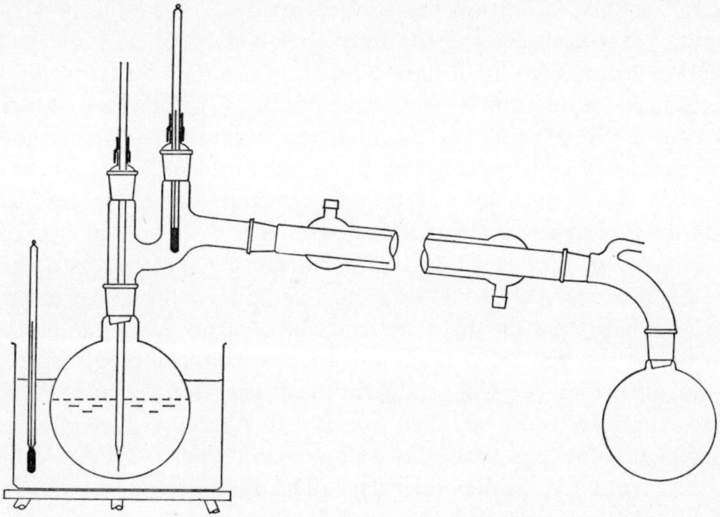

FIG. 5.15. Vacuum distillation apparatus.

boiler flask at the end of the distillation, liquid is forced up into the capillary: this usually cannot be cleared by re-evacuation, and a new one should be fitted before resuming. The drawn-out capillary may be sealed into the distillation system as an extension to a long ground-glass joint, by sleeving it with a short length of pressure tubing or by use of a rubber bung (Fig. 5.15). It is sometimes recommended that a capillary that is too coarse should be closed at the external end by a piece of pressure tube sealed with a screw clip. This is not really satisfactory, and it is much better to replace a faulty capillary.

Condensers. The choice of condensers must always be related to the temperature of the distillation; at reduced pressures, distillation temperatures tend to be low, and efficient condensers are therefore required. Water-cooled condensers are suitable in many cases; if the simple type of Liebig condenser is inadequate, increased efficiency can be achieved with the greater cooling area provided by a multi-surface type, but this is usually suitable only for such operations as solvent removal. The objection to multi-surface condensers for fractional distillation is that they tend to accumulate small static pockets of distillate that can cause serious contamination of successive fractions. If for this or other reasons a simple water-cooled condenser system is inadequate, the efficiency can be augmented by cooling the receiver in iced water, or even in a bath of solid carbon dioxide and alcohol or acetone. A cooled receiver system must be assembled with special care to avoid the possibility of loss of uncondensed material into the pump. The receiver should be almost completely immersed in the cooling bath, and the distillate from the condenser led well down so that it can achieve maximum contact with the cooled surface. Sometimes distillate will solidify at the outlet from the condenser, and great care must be taken to avoid this because it may cause complete blocking with the accompanying risk of dangerous pressure build-up in the boiler flask. The usual precaution against this trouble is to avoid the use of narrow bore tubes for the passage of liquid or vapour into the cooled receiver. For specially difficult conditions it may be necessary to use a pump to circulate cooling liquid at very low temperatures through the condenser, but this is unlikely to be needed by first-year students.

FRACTIONAL DISTILLATION AT REDUCED PRESSURE

Fractionating Columns

The choice of fractionating column for reduced pressure distillations is limited. At atmospheric pressure, a back pressure (resistance to vapour flow) of a few millimetres of mercury is tolerable, but it is much more serious when the distillation is being carried out at reduced pressure. A packed column of any kind is almost useless for distillations at reduced pressure, although at about 15 mm Hg a unit packed with Raschig or Fenske rings may be reasonably satisfactory if the requirements are not too exacting. The most useful general purpose column for reduced pressure distillations is the Vigreux. This unpacked column (Fig. 5.7) is easy to make and to operate, and is probably the best for general purposes.

Receivers

If a column is being used it is especially important to be able to collect successive fractions without interrupting the distillation, because equilibration, on which satisfactory operation depends, is a slow process, and when achieved, should not be upset by any major change in the operating conditions.

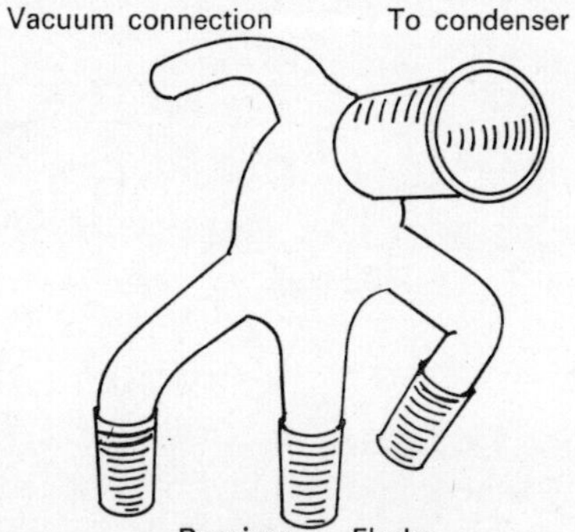

FIG. 5.16. Vacuum distillation receiver "pig".

For small quantities of distillate, a "pig" (Fig. 5.16) is used. This can be turned about the ground-glass joint, by which it is attached to the condenser, to enable successive fractions to fall into separate flasks. For best results, the condenser should have a lip extending beyond the ground glass joint, so that the drops of the distillate fall directly into the receiver flask. This system usually works well for collection of fractions up to about 100 ml in a 250-ml flask. It is difficult to accommodate large receiver flasks on this device.

For larger distillations, a fraction-changing device such as the Kon rectangle, is more suitable (Fig. 5.17). The main distillate

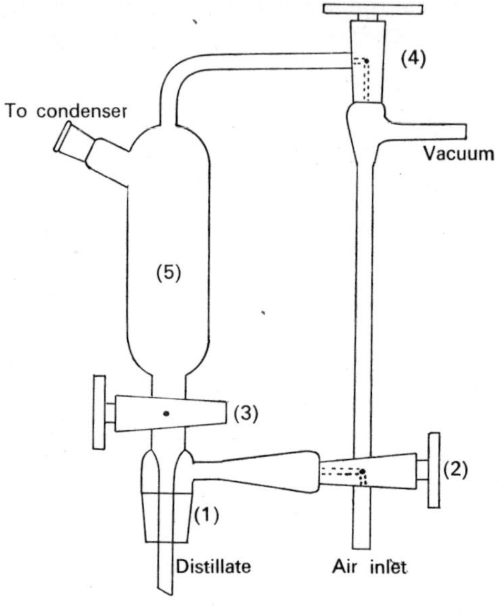

FIG. 5.17. Kon rectangle.

receiver flask attached at (1) may be opened to atmospheric pressure via the three-way tap (2), while the reduced pressure in the rest of the system is maintained by closing tap (3). When the receiver has been changed, tap (4) is closed to isolate the main evacuated system, while tap (2) is opened so as to evacuate the receiver flask; then taps (3) and (4) are opened again, and the distillate which has meanwhile accumulated in the intermediate receiver (5) runs into the receiver flask.

The Kon rectangle has some disadvantages:

(1) High vacua are difficult to obtain, because of inevitable leakage through the large number of taps.
(2) There is some contamination of different fractions in the intermediate receiver.
(3) Volatile distillates may be vaporized from the receiver, because it is not practicable to cool the rectangle.

Thus, although the principle of the Kon rectangle is applicable to any distillation at reduced pressure, in practice it is only used for operations which are too large to be done conveniently with the "pig" receiver system.

Azeotropes

In some cases it is not possible to achieve by distillation a complete separation of the volatile components of a liquid mixture, even though the boiling points are widely different. This condition can arise when two or more of the components form an azeotrope. Azeotropes or constant boiling mixtures are quite commonly encountered and special techniques are required for their separation; simple distillation is not sufficient, however efficient the equipment used. Binary (two-component) azeotropes are exemplified by benzene and water; the individual components boil at 80° and 100°C, respectively, but together they form an azeotrope containing about 7 per cent of water, and boiling at about 69°C. If, therefore, a mixture of water and benzene is heated, the temperature will first rise to 69°C when the azeotrope will start to distil, and this will continue to come over while the two components remain in the boiler; obviously if the azeotrope distillation results in complete depletion of one of the components, continued heating will result in the residual one—water or benzene—coming over as a pure material.

Sometimes ternary (three-component) azeotropes are formed; one example is benzene (74·1 per cent), ethanol (18·5 per cent) and water (7·4 per cent); b.p. 64·6°C.

The most important binary azeotropes encountered in elementary work are those where water is one of the components; the following table indicates the composition and boiling points of some of these:

Non-aqueous component (b.p.°C)		Azeotrope with water	
		b.p.°C	% Water
Chloroform	(61·2)	56·3	3·0
Trichloroethylene	(121·0)	88·5	17·2
Ethylene dichloride	(84·0)	71·6	8·2
Ethanol	(78·5)	78·2	4·4
Benzene	(80·1)	69·4	8·9
Toluene	(110·6)	85·0	20·2

Azeotrope formation can be very effectively applied for the removal of water. For example, benzene as supplied to the laboratory may be saturated with water. This may be removed by desiccating agents such as calcium chloride (see p. 160), but drying by azeotropic distillation can be more convenient, especially if the benzene is to be distilled in any case. Benzene will dissolve about 1 per cent of water at room temperature; if this wet benzene is distilled, the azeotrope containing 8·9 per cent of water will first come over, and from these figures it is clear that all of the water will be removed from the boiler contents in the first 9% (approx.) of distillate. It will be seen that the vapour of the azeotrope contains more water than can be dissolved in the benzene at room temperature, and therefore the condensate will separate into two layers. This phenomenon can be utilized for the continuous removal of water by azeotropic distillation using a device for returning the non-aqueous components of the distillate to the boiler flask. The Dean and Stark apparatus may be used for this purpose, but the design must be adapted to suit the density of the non-aqueous component of the azeotrope, so that either the upper layer (e.g. benzene) or the lower (e.g. chloroform) is returned to the boiler flask. By calibration of the apparatus to measure the volume of water separated, the moisture content of a weight sample of substance dissolved or suspended in the benzene (or other azeotrope former) may be determined.

Extraction

In the present context, "extraction" refers to the use of solvents, to dissolve selectively one or more of the components of a mixture. The term "solvent extraction" may be used for this type of operation, and in the present discussion, the process will be considered from the viewpoint of a liquid solvent being used to extract components from solids, liquids or gases.

Solid–Liquid Extraction

The treatment of a solid containing two or more components with a liquid which will act as a selective solvent for one, but leaving at least one undissolved, is very common; the process is exemplified by the purification of a crystalline compound contaminated with charcoal (see p. 130). The contaminated material is boiled with a suitable solvent, filtered to separate the solution from the charcoal, which remains undissolved, and the pure solute isolated from the hot solution either by cooling or by evaporating or distilling the solvent.

The main problem in this type of extraction is the choice of the solvent; it must show the maximum selectivity with respect to the various components of the mixture. Where there are only two main components, frequently only one lends itself to easy solvent extraction, e.g. with natural products such as plant or animal tissue, where the obvious approach is to use a solvent to dissolve out the desired material which may be present in only very small concentration. If solvents are readily available for all of the components, the choice must depend on many other factors; for example, if naphthalene contaminated with a small proportion of potassium chloride is to be purified, it would be possible in principle, either to dissolve out the impurity with water, or the naphthalene with alcohol or ether. In this particular case it would be better to dissolve the naphthalene in the minimum of hot solvent, filter the hot solution to remove the undissolved potassium chloride and then cool the filtrate to obtain a crystalline deposit of pure naphthalene. In this example it has been possible quite easily to choose a solvent that is highly selective in dissolving out the desired constituent; moreover, the organic material can readily be isolated by allowing it to crystallize from the extract. This use of the solvent for the combined functions of extraction and recrystallization is obviously very advantageous, but it is not always possible: the solvent should always be chosen for its suitability for the extraction itself. In many cases, the best choice for extraction may well be one where the solvent properties are too great to permit easy recrystallization: ether, for example, is frequently useful, but it is not usually a good recrystallization solvent (see p. 127). Having done the separation by extraction, there is no difficulty in distilling off the solvent, and then recrystallizing the residue from another solvent as a separate operation.

The technique of extraction is simple if a solvent is available which will readily dissolve one or more components and lead to an effective separation. The solid should be crushed as finely as possible to expose the maximum surface to the action of the solvent. It is boiled with the solvent under reflux for a short time (10 minutes may be sufficient), and the hot mixture is then filtered. The solid remaining on the filter is washed with a fresh supply of the same heated solvent and the filtrates combined. If the solid is apt to separate from the hot extract on cooling, special care must be taken in carrying out the filtration; the filter itself must be pre-heated and the operation carried out by the method described on p. 147.

In all extraction operations, particularly those where the solid is only slightly soluble in the chosen solvent, it is much more efficacious to extract several times each with a small quantity, than to use the same total volume for one extraction. This applies to the process of washing a solid free from entrained solution just described. The principle is particularly important when the selected solvent does not readily dissolve the material to be extracted. In this case, continuous extraction is used.

Continuous Extraction (Solid–Liquid)

If the required component is only slightly soluble in the chosen solvent, several extractions, each with a relatively small volume of solvent, are usually necessary, and a continuous process using the Soxhlet method is the most satisfactory. This method is commonly used for the extraction of certain natural products from plant or animal tissue, but it is not usually encountered in student work, so it will not be described here.

Liquid–Liquid Extraction

Here, as for solid–liquid extractions, several extractions, each with a relatively small volume of solvent, will be much more effective than the same amount used for one single operation. This is a result of the phenomenon of partition coefficient. Consider a solute, X, soluble in two mutually immiscible solvents, A and B: if a solution of X in A is shaken with B and left to settle, the ratio of concentration of X in A and in B is constant, regardless of the quantities present. The ratio is dependent only on the temperature and the nature of A, B and X. Theoretically, therefore, it is impossible to remove all of the dissolved X from A, regardless of the number of operations carried out, but in practice the extraction can be made virtually complete. Each extraction should be done using only enough solvent B to give a clearly defined layer above or below A. The amount of solvent B required to give two distinct layers may be surprisingly large, because of its solubility in A.

These principles apply to extraction from a system that is initially homogeneous, such as aqueous iodine solution, from which the iodine can be extracted with a solvent such as ether or chloroform, but with slight modifications they also apply to dispersions of liquid in a liquid, and to suspensions of solid particles in a liquid.

In one of the simplest and most commonly encountered heterogeneous systems, a reaction product is obtained in the form of a finely dispersed oil in presence of a relatively large amount of water, which will probably contain other unwanted materials in solution. This type of mixture may be obtained, for example, when a reaction product from an esterification is poured into a large amount of water. The ester is not significantly soluble in water but the other components, alcohol and sulphuric acid, are. Sometimes the oily phase may be present in sufficient quantity to allow it to separate as a clear, well-defined layer which can be removed with the use of a separating funnel, but in most cases at least some of the organic material tends to remain dispersed in the water and solvent extraction must be employed. If the specific gravities of the two liquid phases are similar, there is no gravitational effect to promote separation, and this is one of the factors that commonly causes the non-aqueous component to remain dispersed.

Choice of Solvents

The extracting solvent must be chosen to meet the following conditions:

(a) Maximum solubility for the material to be extracted.
(b) Immiscibility or minimum solubility in the solution.
(c) Chemical compatibility with all the components of the mixture.
(d) Ease of removal after the extraction.
(e) Specific gravity significantly different from the solution.
(f) Ready availability and low price.
(g) Ease of handling, and minimum toxic, fire or corrosive hazards.

Solvents commonly used for extraction of organic materials from aqueous mixtures are:

(a) diethyl ether or diisopropyl ether,
(b) benzene, petrol or other hydrocarbons,

(c) chlorinated hydrocarbons, e.g. chloroform, dichloromethane, etc.

No single one of these is suitable in all cases but diethyl ether is the most commonly used extraction solvent for elementary purposes. Its low boiling point makes its subsequent removal very easy, but the flash point is correspondingly low, so that the fire hazard is rather more pronounced than in many other solvents. Sometimes the mutual solubility of ether and water (7·5 per cent w/v ether in water, and 1 per cent water in ether, at room temperature) is a disadvantage, and because of this diisopropyl ether is a more suitable solvent in many cases; it is quite commonly used in large scale operations, where the lower volatility provides a considerable economic advantage because it reduces losses of solvent by evaporation and makes subsequent recovery more efficient. The use of any type of ether as an extraction solvent introduces a major potential hazard because of peroxide formation. These ether peroxides can be violently explosive, and great care must be taken to remove them before attempting to distil the ether.

Hydrocarbons are commonly used as extraction solvents. Benzene is apt to form emulsions when shaken with aqueous solutions, especially under alkaline conditions, and it is frequently difficult to obtain a clear interface between the aqueous and the non-aqueous phase. There are two possible ways of improving this separation. One is to add an electrolyte, such as sodium chloride or ammonium sulphate (note that the latter is not compatible with an alkaline solution), to nearly saturation concentrations. The electrolyte discourages emulsion formation by its effect on the electric charge of suspended particles or droplets and the dissolved salt increases the density of the aqueous phase, thus improving the gravitational effect on separation. Emulsion formation is also encouraged by the presence of small particles of suspended solids, which act as nuclei: to minimize emulsion formation the solid particles should be removed by filtration. In many cases, the amount of solid present is very small, and the particle size may also be small, so a small amount of kieselguhr or other filter aid should be added to facilitate removal of this very fine semi-colloidal material (see p. 140).

Further objections to the use of benzene, apart from the tendency to emulsion formation, are its high boiling point leading to difficulty in subsequent removal by distillation, and its toxicity (see p. 26).

One of the considerable advantages of benzene as an extraction solvent is that it forms an azeotropic mixture with water. It is therefore not usually necessary to dry a benzene extract as a separate operation. The solubility of water in benzene is low (about 1 per cent), and the small amount of dissolved water is removed as the azeotrope when the extract is distilled to remove the solvent (see p. 114).

The chlorinated hydrocarbons have many advantages and are especially useful when the material being extracted has a specific gravity greater than 1·0. Thus, in the preparation of an alkyl halide, such as butyl iodide (sp.gr. 3·5), the extraction of this material from its aqueous suspension in water is best done with a high density solvent, such as chloroform, which will augment the tendency of the butyl iodide to form a lower layer. Although ether may be used for the extraction, certain proportions of ether and the butyl iodide will have a specific gravity of about 1·0, and so will tend to remain dispersed in the aqueous phase, because of the absence of gravitational promotion of separation. If in spite of this, ether or some other low-density solvent is the most suitable for extraction it is necessary to use a large amount, so that the specific gravity of the resulting solution is sufficiently less than 1·0 to encourage its clear separation as an upper layer.

The use of a high density extraction solvent offers certain manipulative advantages when repeated extractions of an aqueous system are required. Because of the density difference, the solvent forms a lower layer in a separating funnel, so that after its removal the aqueous phase remains, and can immediately be treated with another portion of the extraction solvent. Thus it is possible to carry out a succession of extractions, each with a small volume of the solvent, without the complication of having to remove an upper layer each time. The chlorinated solvents, like benzene, form azeotropes with water (see p. 114) so that the drying of these extracts is facilitated. However, like benzene, the chlorinated solvents are rather toxic, and special care is required if they are being used over long periods of time. On the other hand, they are non-inflammable, and correspondingly safer to use than ether, benzene or other hydrocarbons.

Extraction of Suspensions

For the extraction of suspensions in water, one single extraction may be sufficient to give a well-defined upper or lower layer. It should not be difficult to see why this is the case. The fact that the non-aqueous material is in suspension, even though only a small amount of it may be present, is an indication that the solubility of the suspended material in the water is very low. The extraction solvent merely increases the volume of the non-aqueous phase and possibly changes the specific gravity so that a separate, well-defined layer is formed. In many cases the amount of organic material remaining in true solution in the aqueous phase will be negligible after one extraction, but obviously further extractions may be done if they are considered necessary. A water-miscible solvent such as ethanol may significantly increase the solubility of the suspended organic material in the aqueous phase; the extraction of an ester from an esterification reaction is an example of this, and for this system, multiple extraction is necessary.

Technique and Apparatus

Separating funnels, which are also used for the dropwise addition of a liquid to a reaction mixture, are either cylindrical or pear-shaped; for extractions, the latter are more convenient, because a small volume of liquid forms a relatively deep layer at the bottom of a pear-shaped container.

Efficient extraction depends upon very vigorous shaking to

provide maximum contact between the two immiscible liquid phases, and the funnel must not be overfilled. For most purposes it should be about half to three-quarters full, when all the materials have been added. The funnel must be provided with a well-fitting stopper, preferably ground glass, and the stop-cock suitably lubricated before use.

After adding all the liquid material, the funnel is stoppered, and shaken very vigorously about three or four times, each for approximately 2 seconds. The funnel should be held upside down and the stopper held in firmly during shaking. When shaking a mixture of water and an immiscible volatile liquid, there is a tendency to build up a positive pressure in the closed separating funnel, especially when ether is present. This pressure must be vented between each shaking operation by opening the stopcock momentarily while the funnel is still upside down. This is much more convenient than working with the funnel the other way up and venting by removing the stopper. This venting operation is necessary at all times, but it becomes especially important if a gas is being liberated in the funnel, as it is, for example, if an acidic solution is being neutralized with a carbonate.

When the two liquid phases stubbornly refuse to separate, it is sometimes advantageous to rotate the funnel slowly along its main longitudinal axis, while holding it at an angle of about 45° from the vertical. This is frequently sufficient, but if difficulty persists, some electrolyte (sodium chloride, sodium sulphate or ammonium sulphate) may be added, sufficient almost to saturate the aqueous phase; for most purposes the quantity of salt added can be about 25 to 30 g for 100 ml of aqueous phase; if these efforts fail, filtration of the whole system may be necessary, using a suitable filter aid, such as kieselguhr, and filtering under vacuum through a small Buchner funnel. The filter bed must be washed through at least once with the extraction solvent and with water as well.

Having run off the lower layer through the stop-cock, the remaining liquid (that was originally the upper layer) should be removed by pouring from the mouth of the funnel; this avoids the contamination of the upper layer that would occur if it were also passed through the stop-cock.

On the question of refinement of technique, it is necessary to consider the problem of how to deal with the small amount of liquid that is held in the plug of the stop-cock when it has been turned off after removing the lower layer. It is sometimes difficult to decide just when to stop the separation between the two phases as the interface approaches the stop-cock. The answer can be arrived at quite easily by consideration of the circumstances: if the lower layer is to be discarded (e.g. when washing an ether extract with water), it would be unwise to allow the upper layer to fill the bore in the plug of the stop-cock, and the separation should therefore be continued until the interface just reaches the upper end of the stop-cock bore. In this way, when turning off or when proceeding with another separation, there is no loss of the upper layer containing the desired material. If, however, the upper layer is the one to be discarded, the lower phase should be allowed to run through the stop-cock until the bore in the plug just becomes filled with the upper layer. In some separations, especially if benzene is present, the interface between the two layers is not sharp because of the presence of an intermediate cloudy phase which is an emulsion of the two immiscible liquids. Very often this emulsion layer is not enough to warrant the special treatment mentioned above, but it is still sufficient to interfere with the separation of the two main layers. In most cases, the best rule to apply, if several consecutive separations are involved, such as in repeated washing of a benzene extract, is to leave the small amount of emulsion each time in the layer that still requires further treatment (i.e. in the benzene layer, in the example just quoted). In the final separation, the emulsion can be run off with the layer to be discarded.

When the two liquids in the funnel are left to separate, it should be supported on a retort stand ring; sometimes a tripod stand is suitable. To minimize risk of breakage of the funnel, the iron support should be padded with rubber tubing or a winding of asbestos cord. Supporting a very small separating funnel (*ca.*

50 ml) may present a problem because the stop-cock may be too large to pass easily through a suitable retort stand ring. In this case a section should be cut away from the ring to permit the stop-cock to pass, without danger of knocking it and losing the contents of the funnel (Fig. 5.18). The funnel should usually be left stoppered while waiting for liquid separation to occur; under these conditions, there is always a possibility of pressure building up, and causing the plug of the stop-cock to leak. To prevent total

FIG. 5.18. Retort stand ring modified to support separating funnel.

loss of the contents, it should be made a regular practice to place beneath the stem of the funnel a beaker large enough to accommodate all of the liquid contents. In some cases, pressure in the funnel can cause the stop-cock plug to be ejected violently resulting in a sideways gush of liquid. Because of this, the funnel must be so positioned that its stop-cock is well below the top edge of the protective beaker (Fig. 5.19). This is not always easy if the funnel has a long stem; for separations (as distinct from use as a *dropping* funnel) the stem serves no useful purpose, and considerable advantage results from cutting it off about 1 or 2 cm below the stop-cock. To facilitate drainage from the stem, its end should be cut off at an angle of about 45°. This can be done quite easily with

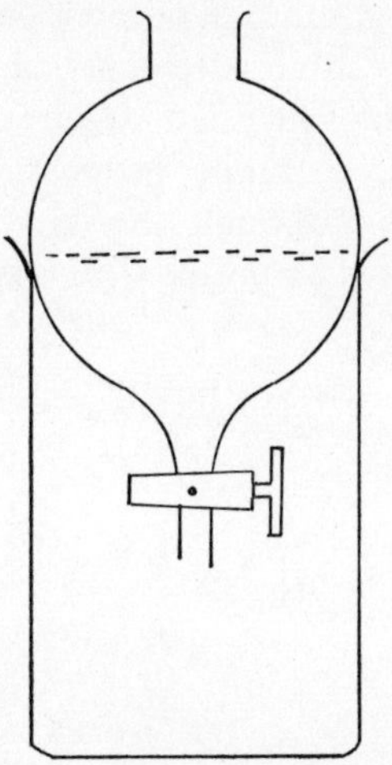

FIG. 5.19. Support for separating funnel.

a wire gauze which is brushed with a short chopping motion across the end of the stem after it has been cut off with a file in the usual way.

Recrystallization

GENERAL PRINCIPLES

This method of purification is used whenever possible and works very well for crystalline substances with melting points above about 40°. Special techniques are necessary for solids with lower melting points. For a proper understanding of the principles and applications, it must always be remembered that recrystallization depends simply upon the phenomenon of solubility and its variation with temperature, as usually represented by a solubility curve.

For a successful recrystallization the solubility of the solid in the solvent should increase rapidly with increasing temperature (i.e. the solubility curve should have a steep gradient (Fig. 5.20(a)). If a hot, near-saturated solution of the impure substance is cooled, the solubility of the solute decreases, and the pure substance crystallizes and can be separated by filtration. The impurity, by definition, represents a minor proportion of the total crude substance; therefore, although the hot solution prepared for crystallization is almost saturated with respect to the substance to

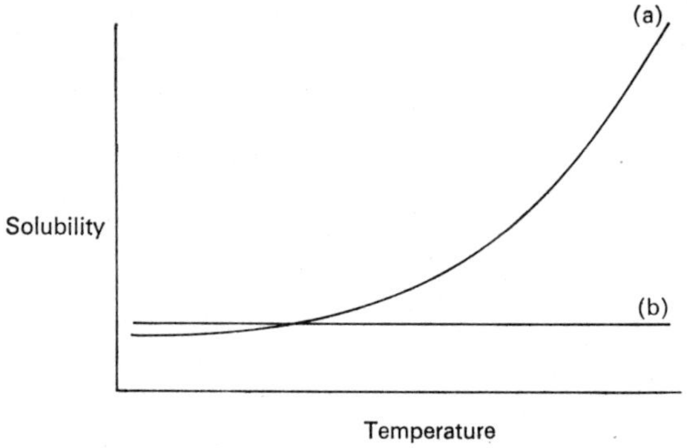

FIG. 5.20. Solubility curves.

be purified, it is not saturated with the impurity, which, on cooling, therefore, remains in solution. The crystals can be separated from the liquid by filtration, which should be carried out using vacuum to remove as much liquid as possible from the solid remaining on the filter, but even so, these crystals will still be wet with the liquid containing the impurities in solution. The crystals must therefore be washed with clean cold solvent of exactly the same composition as that used for making the original solution to remove all the solution adhering to the crystals, which may then be dried. If one recrystallization is insufficient to purify the substance, the whole operation may be repeated.

PREPARATION OF HOT SOLUTION

Choice of Solvent

The solvent for a recrystallization must be chosen to meet the following requirements:

1. The solvent must be chemically compatible with the solute. Thus a carboxylic acid cannot be crystallized from a basic solvent, such as pyridine. In general, the solvents used are relatively unreactive materials, such as water, alcohols or hydrocarbons.

2. The solute must be appreciably more soluble in the hot solvent than in the cold, i.e. the solubility curve for the solute must be steep. For example, sodium chloride cannot be satisfactorily crystallized from water, because the solubility curve is practically parallel to the temperature axis (Fig. 5.20(b)).

3. The solute must not be too soluble; if it is, the hot saturated solution will be very concentrated and have relatively small volume, and the subsequent transfer operation to the filter will be difficult. Also, excessive losses will occur when the crystals are washed. On the other hand, if the solute is not very soluble, large volumes of the solvent must be used and this is inconvenient. As a general guide, the ratio by weight of solute to solvent in the hot saturated solution should be between about 1:10 and 1:5.

4. A solvent, such as diethyl ether, with a low boiling point (about 40°C) is not desirable because the difference between the boiling point and the temperature at which the crystals would normally be collected (usually about 5–10°C) is small, and even if the solubility curve for the solute in question is steep the difference in solubilities over this range will be small so that the recovery of solid on cooling will be poor. Also, the solvent is easily lost by evaporation.

5. The use of solvents with a very high boiling point (above about 120°C) is also undesirable, even if the solubility curve is steep; the solute may separate from the cooling solvent above the melting point of the former, which will thus be obtained initially as an oil and, furthermore, drying the crystals by removal of excess solvent is more difficult. The difficulty of drying the crystals

may, however, be overcome, by rinsing them with a more volatile liquid, miscible with the high boiling point solvent, or by using a vacuum desiccator or a vacuum oven.

The choice of solvent to be used in a recrystallization may be determined on the basis that "like dissolves like". Thus a polar compound, such as a carboxylic acid, is likely to dissolve in polar solvents, such as water, acetic acid or some alcohols. The non-polar, highly covalent materials, such as hydrocarbons, will dissolve more readily in non-polar solvents, such as benzene and other hydrocarbons. The following solvents, listed in decreasing order of polarity, are commonly used for crystallization purposes: water, acetic acid, ethanol, benzene, petroleum ether (light petroleum).

In assessing the suitability of a solvent it is a common error to reject a potentially suitable one because solution does not occur immediately on adding the solid to the boiling liquid. The process of dissolution is relatively slow, and solubility can only be properly assessed by boiling the solvent for at least 30 seconds (and sometimes much longer) in contact with the solid. The rate of dissolution can be increased by having the particle size of the solid as small as possible, and by stirring the mixture. This becomes increasingly important when working with large quantities, or when the solubility is very low; under these conditions the solvent system should be boiled under reflux (to avoid loss) with mechanical stirring for as long as is necessary.

Mixed Solvents

It is not always possible to select a single solvent which is entirely suitable, and mixtures are sometimes needed. In such mixed solvent systems one component is readily able to dissolve the solute, while the other has poor solvent properties; by varying the proportions of the two components, it is possible to obtain practically any desired solvent properties. It is essential that the two solvents in question should be mutually miscible in all proportions, and throughout the temperature range to be employed:

they must also be chemically compatible, both with themselves and with the solute. Commonly used mixtures are ethanol and water, ethanol and benzene, or benzene and petroleum ether. It is an advantage to select solvents having similar boiling points, which should not be too high or too low. For most recrystallizations, the boiling point of the solvent should be about 100°C.

Preparation of solution using mixed solvents. It is usually not possible to predict the relative proportions of the two mixed solvents required, but it is essential to determine this, because the harvested crystals must be washed at least once with a liquid of exactly the same composition as that used for preparing the hot solution. If the crystals are washed with a mixture too rich in the component in which they are very soluble, excessive amounts will be dissolved away; if, on the other hand, the wash solvent is too rich in the component which does not readily dissolve the solid, the mother liquor originally wetting the crystals will be made to precipitate some of the dissolved impurities, and the surface of the pure solid will be contaminated.

The best procedure to follow for preparing a solution using mixed solvents is as follows. Let X be the solid to be recrystallized, readily soluble in A, but much less so in B; liquids A and B are miscible. Two graduated cylinders containing known volumes of A and B respectively, are used as initial supply of solvent. A weighed quantity of X is treated with just sufficient A to give complete solution at its boiling point. Solvent B is then added to the hot solution, until the latter becomes cloudy because of deposition of some X. The addition of B will have caused some cooling of the hot solution, which is therefore re-heated to the boiling point; this heating increases solubility, and the solution again becomes clear. More B is added until clouding just occurs again, and the mixture re-heated. This procedure is repeated until a stage is reached where re-heating just fails to make the cloudy solution completely clear. This indicates that a slight excess of B has been added, and at this point the boiling, slightly cloudy solution is treated with a small quantity of solvent *A*, just sufficient to give a clear solution at the boiling point. The amounts of *A* and *B* taken from the measuring cylinders should be noted at this stage. The solution, on cooling, should yield crystals of *X*. Sometimes it is found that cooling a hot solution prepared in this way tends to deposit *X* as an oil. This indicates that *X* has separated from the solution above its melting point, and in that case it is usually best to cause separation from the cooling solution to be delayed by adding a little more solvent *A*, re-heating to give a clear solution, and again reducing the temperature.

REMOVING INSOLUBLE IMPURITIES FROM THE HOT SOLUTION

If the hot saturated solution is not entirely clear because of the presence of extraneous, insoluble material, such as particles of glass, dust, etc., it must be filtered (see p. 139). Discrimination is necessary in order to determine whether the solid particles in the hot solution are really insoluble, or merely undissolved because insufficient solvent is present, but with a little practice it is not very difficult to judge.

Apart from suspended insoluble particles in the hot solution, other impurities consisting of high molecular-weight coloured polymers, produced as by-products or possibly as a result of atmospheric oxidation may also be present. These coloured impurities usually do not give true solutions under conditions used for recrystallization, but remain as a colloidal suspension in the hot solvent; they must be removed from the hot solution, to avoid contaminating the crystals as they are deposited. Simple filtration will not remove these impurities because they are colloidal and will either pass unimpeded through the filter paper, or clog its pores, with the result that the hot filtration almost ceases.

In order to remove these colloidal particles by filtration they are first adsorbed on to charcoal or other solid such as kieselguhr which is added to the hot solution, and then removed by filtration (see p. 140).

COOLING THE HOT SOLUTION TO PROMOTE CRYSTALLIZATION

Having removed suspended particles and colloidal impurities from the hot solution, the next procedure is to cool it fairly slowly, without agitation, until the temperature has fallen by about 10–20°C and some crystals have separated. After this the solution should be stirred to keep an even temperature throughout the mixture and to encourage the production of small crystals of regular size. Once crystals have separated, nothing is gained by prolonging the cooling process, which may then be expedited by the use of a bath containing water and/or ice.

The solubility curve of the solute in the saturated solvent may be such that, on very slight cooling, separation of the former occurs as an oil rather than a crystalline solid because the temperature is above its melting point. This is unsatisfactory because the oil usually separates in globules which will almost certainly occlude mother liquor and the impurities it contains; also an oil cannot be separated and washed with the same ease as a crystalline solid. In order to avoid separation of the solute above its melting point, either a solvent is used whose boiling point is well below the melting point of the solid or the hot solution is diluted with more solvent so that saturation is reached and subsequent separation of solid occurs at a temperature below that at which the solute melts.

Supersaturation can sometimes cause difficulty; it is a condition of thermodynamic instability that sometimes occurs when a hot solution is allowed to cool slowly and undisturbed. Under these conditions it is possible for the temperature to fall below that at which the solution becomes saturated, but instead of the separation of crystalline solid to form a second phase required for thermodynamic equilibrium, the liquid remains homogeneous. To induce separation of crystals from such a supersaturated solution, thermodynamic equilibrium must be established, and this can be done infallibly by adding a "seed" crystal of the same composition and form as that which will separate. Sometimes a seed crystal can be produced simply by removing a drop of the hot solution at the end of a glass rod and allowing it to cool. If a seed is not available it is often possible to promote crystallization by vigorous stirring, or by gently scratching the inner surface of the container (below the level of the supersaturated solution) with a glass rod or a spatula. If this is not sufficient it may be that simply the mixture should be left to stand—preferably in a refrigerator; it must always be remembered that deposition of crystals under these conditions may be a surprisingly slow process requiring days or even weeks in some cases. As a general rule, however, the rate at which crystals separate is related to the amount of impurities present, and in most simple preparations where the substance to be purified is only slightly contaminated, the difficulties in inducing crystallization are not likely to be very severe; usually stirring the supersaturated solution is sufficient.

In these simple cases, the supersaturation phenomenon can be turned to advantage; sometimes, by leaving the hot solution to cool slowly, and without disturbance, supersaturation will result in the solute remaining in solution until the temperature has fallen below the melting point of the solid. Separation of the solute as an oil is thus avoided and the crystals can then be made to separate simply by stirring or seeding.

In all cases, the problems of oil separation or of supersaturation cannot arise once some of the solute has separated in the crystalline form; by an application of the principles of the phase rule it should be clear that once the system has been brought to a condition where the liquid and the solid coexist, there is no possibility of further disequilibrium and, therefore, the temperature may be brought down to the desired level, as determined by consultation of the solubility curve, with the aid of external cooling.

The most likely cause of difficulty in obtaining crystals is the use of too much solvent. This should always be avoided, but if necessary it may be removed by distillation. It has already been shown that very often excessive volumes of solvent are added because the solid is not allowed sufficient time to dissolve.

The size of crystals obtained in a crystallization is a variable which is normally quite easily controllable. Slow cooling of a hot solution encourages the formation of large, well-formed crystals; rapid cooling promotes the production of small ones. In elementary work it is a common fallacy to attribute some special merit to large, well-defined crystals. They may be aesthetically rather pleasing, but there are few real advantages, and the material separating may be relatively impure because the slow formation of these large crystals adds to the possibility of entrainment of some of the mother liquors (and hence impurities) within the body of the solid: the slow process necessary to obtain the large crystals may also take an undue amount of time. Excessively fast cooling may produce crystals which are so small that their subsequent isolation by filtration is very slow and the filter may even become completely blocked, but this is not very likely, and very often it is sufficient to cool as rapidly as is convenient; if the resulting crystals prove to be too small, it is not difficult to redissolve them by re-heating the suspension and cooling rather more slowly. Probably the best general policy is to aim to leave the hot saturated solution to cool slowly until the first crystals have formed, and then to cool rapidly, taking care always to stir the mixture thoroughly.

Flat, plate crystals, such as are produced by benzoic acid from water are often difficult to filter, because they tend to form compact impervious layers on the filter, and special care should be taken to recrystallize such materials rapidly, with stirring, so that the crystals formed are as small as possible.

Because the process of recrystallization involves the use of solvents, it is impossible to avoid all losses. However carefully the operation is carried out, some of the desired product must remain in the solvent after collecting the main crop of crystals. To minimize these losses the smallest quantity of solvent should be used, and cooling before harvesting the crystals should be very thorough. Most students fail to appreciate that cooling is a relatively slow process; in the majority of cases it is necessary to use some form of artificial cooling medium, such as iced water. Cooling to at least

10°C is usually desirable, and under some conditions considerably lower temperatures may be required. Time can often be saved by leaving the solution to cool overnight in a refrigerator, having first cooled the mixture to about 30°C in cold water. It is not sufficient to guess the temperature of the cooled mixture; it should always be checked with a thermometer before starting to harvest the crystals.

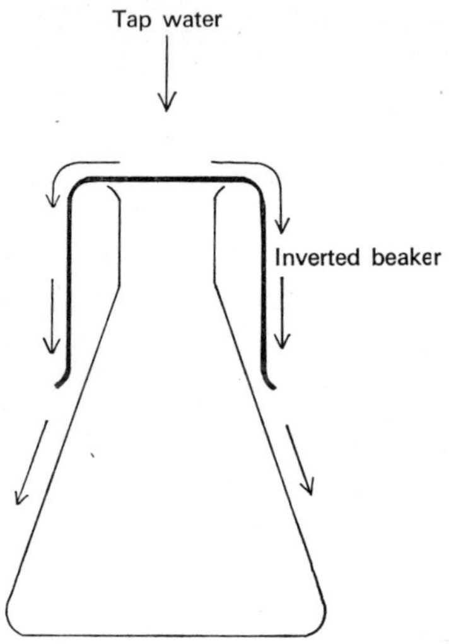

FIG. 5.21. Cooling system for flask.

Because it is frequently necessary to allow solutions obtained in a recrystallization operation to be left to cool for a considerable length of time, it is important to protect them from contamination caused by air-borne particles falling in; whenever practicable the flask or beaker should be covered with a loose lid formed by an inverted beaker or a watch glass (Fig. 5.21). A paper cover is often very satisfactory; it is prepared from a piece of paper (preferably glazed) placed over the neck of the flask and pressed down with the palm of the hand so as to produce a folded skirt.

Precipitation

In some ways, precipitation is analogous to crystallization. Two types of precipitation will be considered here.

An organic compound can be precipitated from solution by adding a liquid miscible with the solvent but which does not dissolve the compound to be separated, e.g. water precipitates benzanilide from a solution in ethanol. This type of precipitation will invariably bring down most of the impurities present, so it is not often used as a method of purification in elementary organic preparations, and its use is not described here in detail.

The second technique of precipitation makes use of the fact that the solubility of organic compounds may change with pH. An acid may be precipitated from an aqueous solution of its salt by adding a strong acid, e.g. benzoic acid is precipitated from sodium benzoate solution by adding hydrochloric acid. Similarly a base may be precipitated from an aqueous solution of one of its salts by a strong base, e.g. naphthylamine is precipitated from a solution of its hydrochloride by addition of sodium hydroxide: the $R—NH_3^+$ ion is much more soluble in water than the base itself.

The techniques of obtaining the free acid or free base from the corresponding aqueous solutions of salts can be exemplified by considering the liberation of naphthylamine in some detail. Suppose the solution contains about 10 per cent of naphthylamine as the hydrochloride, and that there is also an excess of hydrochloric acid. To isolate the base an excess of a stronger base (usually sodium hydroxide) must be added. The minimum amount of sodium hydroxide is that required to neutralize all of the acid present—not only the uncombined acid but also that combined with the amine to form the salt. It is not necessary to go to the trouble of doing an exact neutralization, because an excess of sodium hydroxide does no harm. The usual practice is to add the sodium hydroxide until the whole reaction mixture is strongly alkaline to phenolphthalein. Litmus should not be used because it may indicate an alkaline reaction before all of the amine has been liberated from the salt. When checking pH with a test paper it should not be dropped into the liquid, but a drop removed on a rod and touched on to the indicator paper. In this way contamination of the liquid with the indicator is avoided.

Several refinements of technique must be considered:

(1) If it is required to obtain the base in a high state of purity, it is often advantageous to give the aqueous acid solution a preliminary treatment with charcoal which is then filtered off; this may substantially improve the colour of the amine finally isolated. In any case, the solution of hydrochloride may contain insoluble particles which should be removed by filtration, possibly with the use of a filter aid, before the precipitation is attempted (see p. 140).

(2) It is not desirable to add *solid* sodium hydroxide; a strong aqueous solution is much better, because it permits virtually instantaneous dispersion of the base in the aqueous solution. Sodium and potassium hydroxides are frequently available in the laboratory as alcoholic solutions but these are not suitable for this particular operation, because the liberated base will probably be soluble in the aqueous alcoholic medium, and its isolation will be much more difficult: *aqueous* alkali is usually needed.

The neutralization of an acid with a base is a strongly exothermic process (approximately 13·6 kcal/mole), and therefore the temperature will rise when the amine is being liberated by the alkali. This is not necessarily harmful, but if the base has a low melting point it may separate as an oil; cooling may be necessary to cause separation in the crystalline form. The cooling also minimizes solubility of the base in the water. This need to keep the temperature down conflicts with the recommendation to use concentrated solutions; the larger amount of solvent in a dilute solution permits the heat liberated to be absorbed without causing a severe temperature rise. Sometimes it will be sufficient to cool the amine salt by immersing the container in a water or ice bath, but a much more effective method that is often permissible is the addition of some crushed ice to the amine salt solution. Heat transfer from the solvent to the ice is virtually perfect, and because of the high latent heat of fusion of the ice (80 cal/g) the quantity required for the cooling is relatively small. The ice used in this way should, of course, be clean, and preferably made from distilled water.

There is little point in going to the trouble of cooling the reaction mixture if no attempt is made to know what the temperature is; therefore a thermometer must be used, and stirring is, of course, essential. When excess alkali has been added, the liberated amine is separated from the aqueous liquor. If it is liquid at room temperature, it will probably need to be extracted with a solvent such as ether (see p. 117), but if it is a solid, normal harvesting methods for crystals can be used, bearing in mind that strong alkali is present and it may be better to use a sintered glass filter rather than a Buchner funnel fitted with a paper which may be attacked.

The amine may have separated as a very viscous liquid, or a sticky solid, and filtration is then impossible; the amine will either pass through the filter paper or block its pores and prevent further filtration. Ether extraction may then be necessary, but this may sometimes be avoided by removing the aqueous phase by decantation, or better by suction, using a dip tube, connected via a flask to a vacuum line as shown in Fig. 5.22. Whichever method is used for separating the amine from the aqueous liquors, it must be washed to remove surplus alkali until the washings are neutral to phenolphthalein.

This discussion of the precipitation of an amine from an aqueous solution of its hydrochloride can easily be applied to the parallel case, in which a carboxylic acid is precipitated from an aqueous solution of its sodium salt by addition of a mineral acid. The choice of mineral acid is important. In nearly all cases, it is best to use hydrochloric acid, because it has a high dissociation constant (it is "strong" acid), it is not an oxidizing agent, and it is volatile. The strong acid is needed to liberate the weaker (organic) one from the aqueous solution of its salt. The other two strong acids, nitric and sulphuric, are not suitable for this purpose, because they act as oxidizing agents. This risk may not be particularly great while the acids are in water, but after removing the precipitated material by filtration, the subsequent washing with water may be inadequate and this can lead to a situation where the precipitated organic compound is allowed to dry in presence of traces of the inorganic acid. Hydrochloric acid as a contaminant of the stage may not be harmful, because the process of drying will also remove it by evaporation. If sulphuric acid or nitric acid are present during the drying process, only water volatilizes, so that the organic material will be exposed to heating with increasing

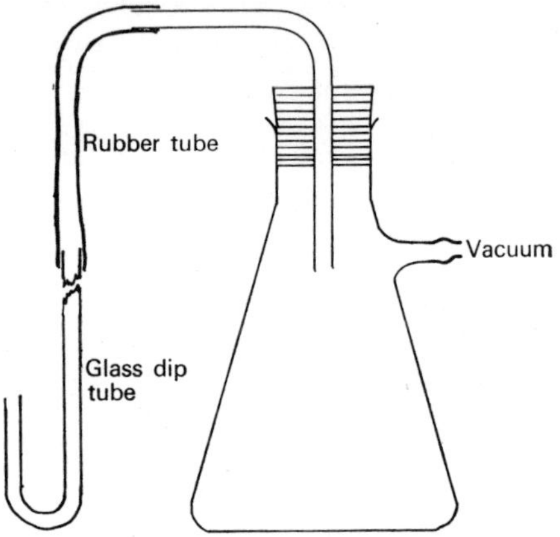

FIG. 5.22. Suction receiver.

concentration of the acid and some damage is likely to occur. In extreme cases, especially if nitric acid is involved, this could be dangerous.

The process of precipitating a solid from a solution is rapid, and therefore the solid usually separates in a very fine form. In some cases, the very fine particles size can be inconvenient, by making the subsequent filtration and washing operation unnecessarily slow. The formation of large particles is encouraged if the precipitating agent is added slowly to the solution at a reasonably high temperature, say 60–70°C, with efficient stirring. When all the precipitating agent has been added, as indicated by the pH, the mixture should be cooled to room temperature before collecting the solid product. This procedure is obviously limited to those cases where the solid being precipitated has a relatively high melting point and is unlikely to suffer any damage from the high temperature of the precipitation. Where cooling is essential, particle size can still be maximized by adding the precipitating agent very slowly, and stirring well.

Filtration

GENERAL PRINCIPLES

Basically, filtration can be regarded as a means of separating the components of a heterogeneous mixture, one of which is of very much larger particle size than the other, by passage through a porous membrane capable of allowing only the smaller particles to pass. In practice, the filtration process is used to remove solid particles from either a liquid or gas; in the present context, only the separation of solids from liquids will be considered.

The efficiency of a separation by filtration is determined by the nature of the porous membrane. If the pores are too large, some of the solid particles may pass through into the liquid: if they are too small, the solid will be retained, but the rate of passage of liquid through the membrane will be unduly slow—in some cases it may be virtually zero. In the majority of cases, the porous membrane is simply a piece of unglazed paper. Filter papers are made in a wide range of porosities, and of different textures to meet specific needs, but they suffer from the limitations of having very low cohesive strength, especially when wet with water, and also, being made of cellulose, they are attacked by some reagents, especially hot concentrated alkali. To meet these more exacting needs, two variants are available. To offset the low cohesive strength of paper, it may be supported in a perforated plate fitted to the funnel; this is found in the Buchner or Hirsch type of funnel. An alternative arrangement is to eliminate the use of paper and replace it with some other porous sheet material such as a textile cloth, but this is only commonly found in large-scale industrial operations. More commonly a funnel is used in which a porous (sintered) glass plate serves as the membrane. The sintered plate is made from powdered glass, heated and compressed under controlled conditions to give a membrane combining the desired porosity with extreme resistance to chemical attack, and adequate mechanical strength.

Using a filter paper, it is sometimes difficult to achieve a satisfactory compromise between a reasonable rate of passage of liquid, and effective retention of the solid particles. This difficulty can arise from two main causes. If the solid particles are very fine, it is difficult to obtain a paper with pores small enough to retain them, while still allowing liquid to pass. Even if the crystals are much larger than the pores of the paper, they can still cause trouble if they are soft enough to compact on the upper surface of the paper and establish a layer that is virtually impervious to the flow of liquid. In either of these situations the cure is provided by the use of a "filter aid". This is a coarse, highly porous powder, such as kieselguhr, which is mixed with the suspension to be filtered, in sufficient quantity to trap the fine particles, or to blend with the slimy ones, and produce an overall solid "mix" that is easily filterable.

In the simplest filtration the force moving the liquid through the pores of the membrane is gravity, but this is not always sufficient, and the flow may be expedited by increasing the pressure difference across the membrane. Usually atmospheric pressure is relied upon to provide the necessary force by evacuating the lower side of the filtration system into which the liquid passes after moving through the filter membrane.

The choice between gravity or vacuum for filtration is determined by the relative proportions of solid and liquid in the mixture to be separated, which component, if any, is to be discarded, and on the temperature of the mixture.

If the separation involves removing a small amount of solid from a relatively large volume of liquid, and if only the liquid is required, it is usually sufficient to use gravity filtration. The suspension is poured into a glass funnel holding a filter paper folded to give a conical or fluted shape (Fig. 5.23). A fluted paper allows faster filtration than paper folded to a conical shape, because in the former there is a greatly increased area unimpeded by contact with the funnel.

Whenever there is a substantial amount of solid to be removed by filtration, gravity alone is not sufficient, because the solid remaining on the filter retains an excessive amount of liquid;

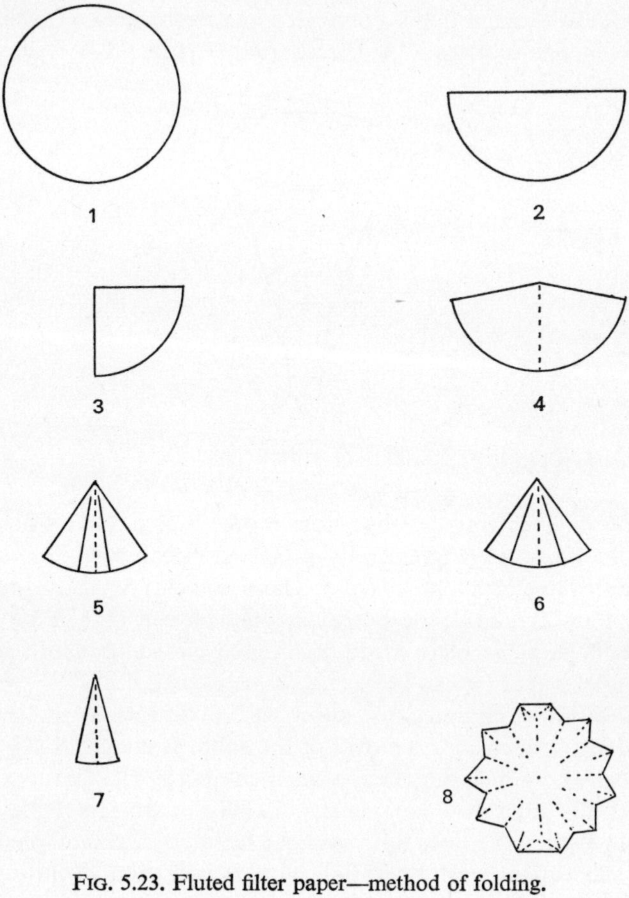

FIG. 5.23. Fluted filter paper—method of folding.

this retention of liquid is only effectively minimized by using vacuum filtration. Vacuum must always be used if the product it is desired to isolate by filtration is the solid component.

SELECTION AND PREPARATION OF THE FILTER

For vacuum filtration the apparatus used is a Buchner flask and funnel—a thick-walled conical vessel provided with a side arm to which the vacuum line is connected and in the neck of which the filter funnel is placed. A Hirsch funnel (Fig. 5.24) is used for

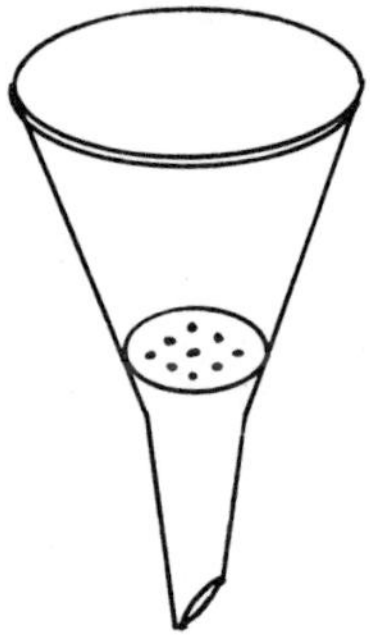

FIG. 5.24. Hirsch funnel.

small-scale filtrations, but for more bulky operations, the Buchner funnel (Fig. 5.23) is preferred because it can accommodate larger amounts of solid. These funnels are usually made of porcelain or glass. The porcelain ones have a perforated plate; funnels with the plate made of sintered glass eliminate the need for filter paper (see p. 143).

The funnel chosen must allow ease of filtration, efficiency of washing, and ease of removal of the solid. It is essential to use a funnel whose filtration area is small enough to enable the solid to cover the filter bed completely, forming a cake at least 3 mm thick. If the cake does not cover the filtration area completely, in the subsequent wash operation, liquors will channel through the uncovered portions of the filter surface, because there the resistance to liquid flow is at a minimum, and the washing of the crystals will be ineffective. The rate of filtration is dependent on the area of the filter, however, and an unduly small filter will not only be very slow, but produce an excessively thick cake which is difficult to wash efficiently. In general, the optimum ratio for diameter to thickness of a filter cake is about 10:1.

Sintered glass funnels are useful because of their extreme resistance to chemical corrosion, and because they eliminate the possibility of contamination of a crystalline product with fibres from a filter paper. The filtration of a hot, strongly acidic solution, which would rapidly damage a paper, can be done through a sintered glass funnel without any difficulty. A sintered glass filter can be re-used, and should therefore be cleaned thoroughly after use (see p. 21). Cleaning is essentially a process of dissolving out the contaminants, so the use of insoluble solids, such as charcoal or kieselguhr, on a sintered glass filter should be avoided. These solids tend to accumulate in the pores of the filter, and may then be virtually impossible to remove.

FILTRATE RECEIVERS

Although Buchner flasks are usually used for this purpose, alternative equipment is sometimes required. If only a very small amount of filtrate is involved, it is best to use a test tube fitted with a side arm, to which the vacuum connection is made. This minimizes subsequent drainage losses when removing filtrate from the receiver. In an alternative arrangement illustrated in Fig. 5.25, a test tube is placed inside the Buchner flask so that filtrate collects in it rather than in the flask itself.

ASSEMBLY OF FILTER FLASK AND FUNNEL

The vacuum seal between the funnel and the neck of the flask may be made by means of a cork or rubber bung, but this is not a very convenient arrangement, and it is much more satisfactory to use a rubber collar or a flat rubber disc as a seal. The disc, made of soft rubber about 3 mm thick, is provided with a central hole, through which is passed the stem of the funnel and it is large enough to cover completely the neck of the filter flask (Fig. 5.25). This seal is quite satisfactory unless there are serious imperfections in the top edge of the neck of the flask. There is a great advantage over a bung that with this rubber gasket the funnel can be seated

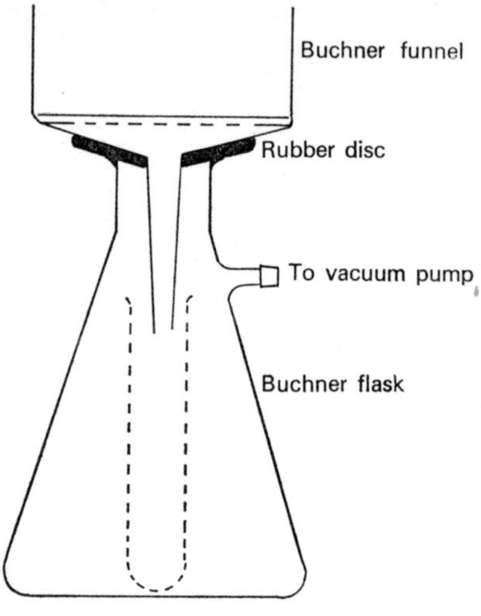

FIG. 5.25. Vacuum filtration assembly.

satisfactorily in a variety of flasks of different neck sizes. In using this type of seal, it is important to remember that as air is first sucked out of the flask, the funnel will tend to be pulled into the neck as the rubber flexes. Sometimes this movement causes the funnel to tilt, and if liquid is present in it, some can be lost by spillage. It is therefore desirable to steady the funnel with the hand as the filter paper is wetted and/or as the vacuum is applied, so that no damage or spillage is caused as the filter settles. On releasing vacuum, the funnel should again be steadied while the rubber disc lifts.

When using the Buchner flask and funnel arrangement, it is important to prevent the filtrate being sucked down the vacuum line. This risk is best avoided by selecting equipment so that the lower tip of the stem of the funnel is well below the level of the side arm on the flask. The end of the funnel stem is usually cut off at an angle of about 45° to the main axis, to facilitate drainage, and, contrary to what might at first seem best, the fitting into the flask should be such as to bring the upper edge of the stem as near as possible to the side arm. One of the advantages of using a disc rather than a bung to seal the funnel into the flask, is that with the former, the funnel is placed much lower in the neck of the flask, and so the risk of sucking filtrate into the side arm is reduced.

A filter paper to be used with a Buchner or Hirsch funnel must be of the correct size. Obviously the diameter of the paper must be sufficient to cover all the perforations on the supporting plate, but it is equally important that the paper should sit flat on the plate, and not be so large as to require folding at the periphery. Filter papers of the appropriate size for the common range of Buchner and Hirsch funnels are normally available, but if not, a larger sheet must be cut to size. The paper should have a diameter about 2 mm less than the internal diameter of the filter plate. In order to trim an oversize paper to fit, it should first be rested on the upper edge of the filter, and then pressed down by brushing the palm of the hand round the perimeter; this produces a circular mark on the paper, which is then cut with scissors to give a circle about 0·5 cm less in diameter.

Before actually transferring material to be separated on to the filter, the paper must be properly bedded down, by wetting it with the same solvent as in the substance to be filtered. If the paper is not wetted, it will be impossible to obtain a satisfactory vacuum seal, and solid may then pass through, underneath the paper, when the filtration is started. If this condition develops it is virtually impossible to cure it, and the only course is to transfer the material collected in the filter flask back to the original container, clean out all the apparatus, and start again.

The filter paper may not bed down properly in the funnel because the plate of the funnel is not truly flat, and in this case the funnel should be rejected.

TYPES OF FILTER PAPER

Filter paper must be fine enough to prevent solid particles passing through with the filtrate, coarse enough to give a reasonable rate of passage of filtrate through it, and mechanically strong enough to withstand the effect of pressure differences and the weight of the material upon it. This point is especially important when the paper is used in the presence of water, when the "wet strength" may be very low.

For most vacuum filtration especially of aqueous solutions, a hardened filter paper, such as Whatman No. 40, 41 or 42 should be used. In this system of labelling grades, the prefix 4 indicates that it is a hardened paper, and the second number indicates the porosity and general filtration characteristics (not the dimensions). For analytical purposes, special grades of filter paper are available, in which the weight of ash after ignition is either negligible, or fairly accurately known.

SINTERED FUNNELS

For most purposes, the mechanical strength of the filter plate of a sintered funnel can be taken for granted; this type of apparatus can be used under vacuum without any special precautions and the only variable of practical importance is the porosity. The sintered plates are made in a range of porosities, which are numbered 1 to 4, the larger number indicating a finer pore size. Nos. 2 or 3 filters are most commonly used. The use of No. 4 filter is usually only justified when dealing with very fine particle sizes because the rate of passage of liquid through the plate is apt to be slow.

Although the porosity of the sintered glass or paper may be a limiting factor for the filtration rate, this is not always the case, and sometimes the shape and the size of the solid particles accumulated on the filter plate are more important. The rate of flow of liquid is a function of the porosity of the filter itself, and also of the compacted bed of solid above it; the overall rate is determined by whichever of these gives the slower flow. Obviously, a bed of very fine particles resting on a relatively coarse filter plate will still prevent rapid flow of liquid, and hence give a very slow rate of filtration. If the solid particles being filtered are very fine, they may enter the pores of a very coarse filter, and cause serious blockage that may impede or even totally stop the liquid flow. Because of this, when dealing with very fine solid particles, a fine filter plate may give a better rate of filtration than a coarser one. Usually the controlling factor is the particle size of the solid, and only by changing that is it possible to make any substantial improvement in the filtration rate. In some cases the permeability of the filter cake to liquid is not so much a function of particle size as crystal shape. Thus, large, flat plate crystals tend to pack down on the filter, and because of their face-to-face contact like slates on a roof (or fish-scales), the rate of flow of liquid through the crystal mass can be surprisingly slow. Ideally, the answer to this problem is to recrystallize under conditions that produce a different crystal shape; this is not easily done in many cases, although sometimes a change of solvent will serve. In the majority of cases, the fact of having to deal with plate crystals must be accepted; the solution should be cooled rapidly and stirred very vigorously, preferably with a mechanically driven stirrer, so as to produce broken small crystals, which can be filtered off with least difficulty.

HOT FILTRATION

Liquids to be filtered while hot are invariably solutions, often saturated with a solute which is being purified. The simplest case is represented by a hot solution being clarified without the use of charcoal, or if charcoal has been added, no filter aid is present and so the amount of solid to be removed is very small. Here, no vacuum is necessary, and the liquid can be filtered through a fluted filter paper supported in a heated conical glass funnel from which the stem has been removed; the presence of a stem on the funnel is undesirable, because the filtrate may cool in it, and cause blockage by depositing crystals.

If a filter aid such as kieselguhr is added to assist the removal of charcoal or other fine insoluble material, the total bulk of solids to be removed is relatively large, and vacuum filtration using a Buchner or similar funnel, is necessary.

Filtration of a hot solution under vacuum can cause the filtrate to boil, and some solvent vapour may be sucked down the vacuum line. Therefore a water pump should always be used for this purpose because this will not be harmed by the vapour; a mechanical pump may be damaged by solvent entering it. Loss of solvent from the filtrate should be compensated by re-heating to boiling, and if the crystals do not completely dissolve, sufficient extra solvent (of the correct composition in the case of mixed solvents) should be added to give a clear solution at the boiling point.

Even if filtration under vacuum is used to remove suspended solids from the hot solution, the residue remaining in the filter will still contain an appreciable amount of occluded liquor which must be washed out immediately with boiling pure solvent if low yields are to be avoided. Delay at this stage may allow the crystals to separate from the liquor retained in the filter cake; a supply of hot wash solvent of the correct composition must therefore be available before starting the main filtration. This should be poured on to the filter just as the liquid in the funnel passes down below the upper surface of the cake of solids. The quantity of wash solvent used should be roughly equal to the bulk volume of the solid cake, and the liquid passing through should be combined with the main filtrate.

If a filtration is very slow, or if traces of charcoal still pass through into the filtrate, in spite of a filter aid, a funnel should be prepared with an extra layer of filter aid, about 3 mm deep, on the filter paper before carrying out the main filtration. This "pre-coat" is applied to the filter paper by making a thin slurry of the filter aid in the clean, boiling solvent and filtering it through the hot filter funnel immediately before carrying out the main filtration of hot solution: the filtrate from the pre-coating operation is discarded before starting the main filtration.

Heating the Funnel

A small glass funnel for use with a fluted paper may be heated by cautiously waving it in a Bunsen burner flame that is neither luminous nor roaring. The relatively poor thermal conductivity of the glass makes it possible to hold the funnel between the finger and thumb at the edge furthest from the stem without discomfort while heating the main part of the glass to a suitable temperature. The glass will shatter if heating is excessive, especially if the funnel is not kept moving while exposed to the flame, but it is sometimes difficult to judge just how much heat is needed. Excessive caution will result in the funnel being too cool for the subsequent filtration, so that premature crystallization may cause blockage. If the funnel is over-heated, apart from the initial risk of breakage, the subsequent addition of the hot liquid to the filter itself may cause the glass to shatter, or solvent may boil off when it comes into contact with the very hot glass, so that premature crystallization occurs. Ideally, the funnel should be at the same temperature as the liquid to be filtered, but the only simple way of judging the temperature of the former is to touch it cautiously. It is useful to remember in this context that body temperature is about 37°C, and most people regard a surface temperature of about 60°C as just about too hot to touch more than momentarily.

This method of heating in a bare flame cannot be used satisfactorily for larger funnels, or for the Buchner or Hirsch type, mainly because these have a stem which is difficult to heat properly by the method described.

If the funnel can be heated in an oven, temperature control presents no difficulty and there is no serious risk of breakage from thermal shock, but the method is relatively slow. The funnel should be left in the oven for at least half an hour, and therefore it is important to anticipate the need for this if waste of time is to be avoided.

Subject to some distinct limitations, one of the best methods of heating a funnel is to use a steam bath. If the liquid to be passed through the hot filter contains a substantial proportion of water, a steam bath may be used to heat the funnel which is placed on the bath, with the stem pointing vertically upwards, so that steam may pass up through it. The heating, which is quite rapid, is continued until steam emerges copiously from the stem, when it can be assumed that the whole of the surface of the funnel has reached a temperature of nearly 100°C, which is adequate for most purposes. Before use the funnel should be withdrawn from the steam bath and shaken vigorously to displace any surplus water, and then used immediately for the filtration. It is difficult to dry a funnel that has been heated in a steam bath, and the method is therefore inapplicable to the filtration of a non-aqueous mixture, or in any other case where harm could come from addition of water.

HARVESTING SOLIDS—TRANSFER TO THE FILTER

The transfer of the crystallized material to the Buchner or Hirsch funnel in order to isolate the crystals requires considerable care. When the funnel and flask have been prepared for use, with the filter paper bedded down (see p. 145), and the vacuum lines connected, the cooled crystallizing mixture is stirred thoroughly and poured as quickly as possible on to the filter.

During the transfer of material to the filter, the funnel should be kept about three-quarters filled with the slurry. The transfer of slurry to the filter is sometimes difficult because the mixture may have acquired a pasty consistency and does not flow or drain completely from its container even when well stirred. The difficulty is minimized if the crystallization has been carried out in a wide-mouthed conical flask, but is at its worst if a narrow-necked spherical vessel has been used. Having transferred as much of the slurry as possible, the operation may be completed by first thinning the residue with some of the filtrate that has already been collected, followed by further cooling if necessary. Because these liquors are already saturated with the solute they can be used repeatedly for this backwashing technique without fear of dissolving more of the solid.

The flask and funnel used for harvesting solids should be clean and dry, before the filtration is started; but if the apparatus is initially wet with water, and the solvent to be collected as filtrate is water soluble, it is permissible to rinse several times with the latter. The main point is that the flask and filter must be either dry or only be wet with solvent of exactly the same composition as that used for the recrystallization. Failure to take this precaution can cause considerable difficulty, and if there is any doubt about the composition of residual liquid in the filter flask, it must be removed before filtration is begun. Consider, for example, the situation if water is present in the filter flask when collecting an alcoholic filtrate. The water could then cause precipitation of both crystals and impurities from the alcoholic solution, and subsequent transfer of contaminated filtrate to the main one, in the backwashing procedure described above, would necessitate complete repurification of the solid. It should be remembered also that even if backwashing is not intended, refiltration may be made necessary by such a mishap as perforation of a filter paper. If the liquid in the filter flask becomes contaminated with air-borne, solid particles when the vacuum is released, the filtrate should be refiltered through a fluted paper before using it for backwashing.

WASHING THE SOLID ON THE FILTER

In spite of the vacuum filtration, the solid will still be wet with mother liquor containing the impurities which must be removed by washing the crystals with clean solvent. The fundamental principle to follow is that once a recrystallization has been done, the first washings must be of the same composition as that of the solvent used for preparing the original hot solution. In order to minimize solubility losses only the smallest quantity of washing solvent should be used, and it must first be cooled to the same temperature as that of the slurry from which the crystals were removed by filtration.

There are two techniques for removing mother liquor from the crystals: the "displacement" method, and the "slurry" wash method.

Displacement Washing

The displacement method is preferred because it uses a minimum of solvent. The mixture of solid and liquid transferred to the filter is sucked down until the liquid level on the filter is just about to move below the surface of the cake; if the rate of filtration is rapid, it may be desirable to interrupt the vacuum supply. Then a further amount of cooled solvent is carefully added, without disturbing the surface of the cake, and this is allowed to suck down until the liquid level is just about level with the solid: the process is then repeated. In each case the quantity of solvent added is such as to give a layer on top of the solid not more than twice the thickness of the cake.

This method can only give satisfactory results if movement of liquid through the cake is evenly distributed, and the latter must therefore be kept free from cracks or from any tendency to pull away from the sides of the filter. For effective displacement washing, it is absolutely essential that the wash liquor should move evenly and steadily through the homogeneous cake in such a way that the contaminating liquors are displaced completely by the layer of clean solvent above. If the vacuum is allowed to suck the solid dry before the next wash is added, cracks are very likely to appear and once this damage to the bed has occurred it cannot be repaired simply by smoothing over the surface with a spatula because the deeper layers of the cake will still have such an irregular texture and consistency as to encourage channelling and uneven distribution of liquid. Once this condition has developed, only slurry washing can be applied.

Slurry Washing

When displacement washing has failed, slurry washing may be employed, although it requires more wash solvent and there is a risk of damaging the filter paper. Slurry washing is preferable to the displacement method only if the fine particle size of the solid makes filtration very slow.

The solid on the filter is left under vacuum long enough to remove as much of the original mother liquor as possible, and the cake is pressed to a firm consistency with a spatula or flat glass stopper. When no more filtrate comes through, the vacuum line is disconnected and just enough of the cooled wash solvent is added to be dispersed thoroughly in the cake, using a spatula and taking care to avoid damage to the filter paper; vacuum is then reapplied. The procedure is repeated, taking care that the cake is pressed down well and the maximum amount of filtrate collected between each wash. The process of dispersing the wash liquor in the solid and subsequent removal by filtration is repeated as many times as necessary. Two or three washes are usually sufficient

If the mother liquor to be removed from the solid is coloured, washing should be continued until the filtrate comes through colourless. This is only applicable if the compound being washed is not itself coloured. If the solid is coloured, the washing solvent will dissolve some of the solid, and so may always come through slightly coloured. If the mother liquor contains acid or alkaline impuritics, the washing can be judged complete when the filtrate is at the same pH as the pure solvent. The test is done by removing the filter funnel and touching the tip of the stem on to a suitable indicator paper.

If, during this washing operation, the filter paper is damaged, when the vacuum is reapplied some of the solid may be sucked with the filtrate through the damaged paper and mixed with the filtrate already accumulated in the filter flask. Recovery of the solid then necessitates refiltration, and the *whole* wash procedure must be repeated. To avoid this, the main filtrate first collected (before the first wash) should be transferred to another container before slurry washing is commenced. Then, if the paper should be damaged, the amount of liquid needing to be refiltered will be very small, and contamination with the impurities in the main filtrate will be avoided.

Change of Washing Solvent

In the preceding discussion, the need for using wash liquors of exactly the same composition as that used in the original crystallization was emphasized. This is essential up to the point where all the impurities have been washed out, and it would be feasible to remove the filter cake and dry it. In some cases, the solvent used will be relatively high-boiling, and the drying operation may be correspondingly difficult; it may then be advantageous to accelerate drying by displacing the original solvent with another one with a lower boiling point. The new wash solvent, however, must be miscible with the solvent system originally used and it must not dissolve the crystals appreciably. Above all else, it must be emphasized that this change of solvent may only be made when all the necessary washing has been done with the original one. If the filtrate originally collected is to be used for subsequent recovery of a second crop of crystals, it must be transferred to another container before the new volatile solvent is added to the crystals in the funnel; adding the new solvent to the filtrate already collected may precipitate impurities, and impede the subsequent procedure for obtaining the second crop of crystals. Since the volatile solvent used for the final wash should have only very low solvent power for the crystals, it should not be worth while attempting to obtain a third crop of crystals from it.

In all cases, after isolating crystals by vacuum filtration, it is important to disconnect the vacuum whenever it is not actually needed. It is a common fault, when the solid has been filtered off and all the washing completed, to leave the filter assembly connected to the vacuum long after the last drops of filtrate have been collected. This is a bad practice because it only serves to extract air-borne particles, which will then contaminate the solid on the filter.

Removal of Filter Cake from the Funnel

The cake should be removed intact from the filter using a thin, fine spatula to cut down the depth of the solid as near to the wall of the funnel as is possible. The funnel is then inverted over a watch glass or a large sheet of glazed paper, and shaken sharply. Once the cake has been detached from the side walls of the funnel, it will usually drop out very easily with the paper still attached to it; it is then easy to peel the paper off the solid.

If the filter cake does not readily detach itself from the plate of the funnel, a spatula should be used as a lever underneath the paper. As a last resort, if these procedures fail, the solids must be scooped from the filter plate by the use of a spatula. Care must be taken not to damage the filter paper; the solid should be removed in thin layers, preserving the paper intact.

Second Crops

It is impossible to obtain truly quantitive yields from a crystallization: after the main crop of crystals has been separated the filtrate (known as "mother-liquor") will be just about saturated with solute, some of which may be recoverable. This solution is now relatively richer in impurities and correspondingly poorer in the desired material. Also, the solution is only just saturated at the lower temperature at which the crystals were originally harvested, so there is too much solvent present to permit crystallization of the liquors as they stand.

In some cases it is possible to obtain a second crop of crystals simply by removing solvent by evaporation or distillation until incipient crystallization at the boiling point is observed. On cooling the concentrated solution as for the main crop, it may be possible to obtain another batch of crystals. These recovered crystals, having been obtained from relatively impure sources, may be more highly contaminated than the main crop, and so need to be recrystallized from clean solvent, in order to obtain them in a reasonable standard of purity. In some cases, concentration of the mother liquors followed by cooling only gives a gummy, dark solid, which does not crystallize; this is mainly because of the presence of coloured polymeric materials, which inhibit crystallization. These can be removed with decolorizing charcoal as described on p. 130. In general, if the mother liquors are very highly coloured, this treatment with charcoal should be done before concentrating them; this will remove most of the impurities and improve the prospects of obtaining a second crop of crystalline material.

Procedures used for drying the solids isolated by the filtration techniques described above are discussed in the next chapter.

CHAPTER 6

DRYING

Drying Solids

The liquid held in the filter cake is removed by evaporation, expedited by placing it in an oven at a temperature giving the maximum rate of drying without causing damage to the solid. The oven temperature must be below the melting point of the solid being dried: it is unwise to try to dry a solid at a temperature higher than 20°C below the melting point. Further limitations may be imposed by the tendency of the solid to decompose or sublime. In all cases, tests should be made to ascertain that the proposed drying temperature will not be excessive.

When it is impracticable to dry a solid at an elevated temperature, a desiccator is used. For absorbing water the usual desiccants are phosphorus pentoxide, concentrated sulphuric acid, fused calcium chloride, or silica gel. If acidic vapours (e.g. acetic acid or hydrochloric acid) are to be absorbed, solid potassium hydroxide is probably best. Calcium chloride will absorb alcohol vapour and ammonia as well as water. For the other common organic solvents, such as ether, benzene, or petroleum ether, a dish of paraffin wax parings is useful as an absorbent.

To increase the rate of drying the desiccator can be used under reduced air pressure, but if this is done, precautions must be taken to minimize the damage resulting from implosion if the desiccator should collapse under the effect of the external pressure; the usual safeguard is to keep the desiccator in a strong wire cage. A vacuum oven combines the benefits of a vacuum desiccator and an oven, and is useful for drying large amounts of thermolabile materials.

It is always important to take special care when opening an evacuated desiccator. The air inlet cock should be opened only very slightly to avoid the possibility of a sudden rush of air causing the dried solid to scatter. Sometimes provision is made for air to be admitted only through a fine capillary tube. It is a useful extra refinement to place a filter paper over the air inlet tube; this remains attached by vacuum, and provides an effective filter for the ingoing air. The paper falls away from the end of the inlet tube when the desiccator is at atmospheric pressure.

The time required to dry a solid is often underestimated. The removal of water is particularly slow because of its high latent heat of evaporation; also, the amount of water or other solvent to be removed is usually underestimated. A cake removed from a Buchner funnel, sucked as free from filtrate as possible, may still contain up to 50 per cent of its total weight as liquid. Drying in a desiccator, even when evacuated, is a slow process, usually taking several hours. Oven drying at elevated temperatures is obviously a more rapid process, but even here the times involved must usually be reckoned in hours rather than minutes.

It is important to see that drying is complete, and in the early stages of gaining experience of organic chemistry yields that are surprisingly high (even more than 100 per cent of theory) can be accounted for by the fact that the product has been weighed before it is dry. The practice of drying to constant weight is just as important in organic preparative chemistry as it is in gravimetric analysis, and provides the only satisfactory evidence that the sample in question has been freed of all solvent. The substance being dried is weighed at intervals of about 30 minutes until two consecutive weights are identical or do not differ by more than a few milligrams.

Drying Liquids

The solubility of water in organic liquids varies from complete miscibility to quite low values, and it is never justifiable to assume without specific evidence that any solvent or reagent is completely dry. This is especially important in carrying out such reactions as the Grignard, where even small concentrations of water (of the order of 10 ppm) in the solvent or reagents will be harmful. Anhydrous solvents may be prepared by removal of water by physical methods (usually distillation), by absorption on desiccants such as calcium chloride, or by causing it to react with sodium.

REMOVAL OF WATER BY PHYSICAL METHODS

Fractional Distillation

If water does not form an azeotrope (see p. 113) with the organic liquid to be dried, drying by distillation is certainly possible, as long as there is a reasonably wide difference in boiling points: a fractionating column may be needed if the boiling point difference is small.

As a general rule, dehydration methods involving fractional distillation are only acceptable if they can be applied simply and without the use of high-precision equipment. Thus, it would be reasonable to dry acetone by fractional distillation, but if the non-aqueous component has a boiling point above 100°C and the concentration of water is small, other methods should be used.

Azeotropic Drying

Benzene, toluene, xylene and the chlorinated hydrocarbons, such as chloroform, can each be effectively dried by distilling out the relatively low boiling azeotrope that it forms with water. (Ethanol can be dried by the formation of a ternary azeotrope with water and benzene, but usually a desiccant is used.) Azeotropic drying is only possible if the concentration of dissolved water in the saturated organic liquid at room temperature is significantly lower than that in the azeotrope at its boiling point. Distillation of this saturated material will first produce the azeotrope—a low boiling fraction, relatively richer in water than the wet solution—which will distil until no water remains: from that point onwards pure dry solvent will distil at a higher temperature. The azeotrope produces a characteristic appearance as it distils; it is relatively richer in water than the saturated solvent at room temperature, so that, as it is cooled by the condenser, some water separates to give a heterogeneous mixture which will initially look milky because of suspended water globules, but may separate into two layers in the receiver. Distillation is therefore continued until the distillate emerging from the condenser is quite clear. In order to judge this accurately, small quantities are collected separately in a test tube. Once the drying has been achieved, it may be sufficient to use the residual liquid in the boiler flask without further distillation: if, however, further distillation is required, the original condenser and other wet fittings should be changed, or sufficient distillate should be allowed to pass over these surfaces to wash them free of water, before the dry component is collected.

REMOVAL OF WATER BY FREEZING

This is not usually a satisfactory drying method, but it is important to remember that if a solvent such as petroleum ether is being used at very low temperatures, such as given by carbon dioxide and acetone (−78°C), any water present will separate out as ice, producing an opalescent effect.

REMOVAL OF WATER BY DESICCANTS

Even where distillation is sufficient to remove most of the water, small traces may still remain, and in all cases where a liquid must be absolutely dry, a final treatment with a drying agent should be applied. Desiccants must also be used for those materials which cannot be dried by distillation.

The drying agent commonly preferred for drying organic liquids is fused calcium chloride, but potassium hydroxide, sodium metal, phosphorus pentoxide, calcium oxide, magnesium sulphate and sodium sulphate are also used. The choice of desiccant must always take account of the possibility of chemical incompatibility; for example, calcium chloride cannot be used for drying any of the alcohols or amines. The drying agent must be insoluble in the organic liquid, and its speed and efficacy depend upon surface contact, which should be as great as possible. In order to promote this contact, the mixture of solvent and drying agent should be shaken periodically. These drying agents are not always very rapid in their action, and solutions should be left in contact with them for a reasonably long time. Small quantities of liquid, drying over calcium chloride, will be ready for use in about 1 hour, but, as far as is possible, stocks of the more commonly required materials, treated with the desiccant, should normally be available. Prolonged contact with the desiccant is not harmful.

There is usually a tendency to use far too much of these desiccants. The drying agent changes its physical form as it removes the water from the solvent being dried. Thus, calcium chloride develops a moist, sticky surface, and the separate granules tend to stick together; pellets or sticks of sodium (or potassium) hydroxide behave in a similar way. Sodium metal, usually used in the form of wire for drying ethers and hydrocarbons, tends to disintegrate and dissolve. These common drying agents therefore provide a clear visual indication when they have come into contact with anything more than the smallest trace of water, and while they retain their original physical form and appearance, they are still capable of exerting their drying capacity and it is unnecessary to add more. However, since the speed of drying is essentially dependent on area of contact between the desiccant and the liquid, use of rather more than the minimum quantity can sometimes be justified in order to obtain an increased rate of water removal.

For most practical purposes for preparing laboratory supplies, fused calcium chloride in the form of granules of about 6–10 mesh can be added to a flask or bottle full of solvent, to provide a thin layer. If stocks already prepared in this way are not available, and immediate supplies of dry solvent are required, two or three times this quantity of calcium chloride, preferably crushed to a coarse powder to increase surface area, would shorten the drying time to about one hour; the mixture should be shaken at frequent intervals. The advantages of powdering the calcium chloride are offset by the necessity of filtering the liquid to be sure it is free from suspended solids. If a small amount of granular calcium chloride is used, it is permissible to decant off the dried liquid, although it should be filtered wherever possible. This filtration is best done through a fluted filter paper.

Sodium metal is often used for drying ethers and hydrocarbons, and is probably a marginally better desiccant than calcium chloride. When sodium is to be used as a desiccant, the usual technique is to extrude the metal through a suitable press to produce a thin wire, about 1 mm diameter, which is passed directly into the bottle containing the solvent: no attempt must ever be made to transfer the wire from one container to another. Winchester quart bottles are commonly used as containers; the bottle should be about three-quarters filled with the liquid to be dried, and sufficient sodium wire added to give a very loose network spread throughout the liquid. For this purpose, about two pieces of sodium metal, each about 1 ml in volume, should be extruded into each bottle.

The drying action of sodium depends upon the ease with which it reacts with water; after adding the metal to the wet solvent, it is essential to leave the container vented to permit the hydrogen produced in this reaction to escape, but the bottle should not be left completely unstoppered, because of the possibility of contamination, and the risk that water may be splashed into it. The best method of protecting these bottles is to close them with a cork or bung through which a freshly filled calcium chloride drying-tube is fitted. It should be remembered that the calcium chloride tends to cake on prolonged exposure to moist air, and become impervious to the passage of gases.

Drying Solutions

The procedure for the drying of solutions is not different in principle from that used for solvents, and can be done either by distillation methods or by the use of desiccants, but the detailed requirements are more exacting. In particular, the question of chemical compatibility must be carefully considered: the drying agent must be compatible (i.e. chemically inert) with respect to the solute as well as to the solvent.

The need to dry a solution commonly arises after a non-aqueous, water-immiscible solvent has been used to extract a component dissolved or suspended in water. Solvents commonly used in these extraction procedures are ether, benzene or other hydrocarbons, or chlorinated solvents such as chloroform: all of them dissolve small quantities of water. These solvents are used because they are relatively inert chemically, and therefore in the majority of cases, the choice of desiccant will probably be determined by the nature of the solute. The solvent used in an extraction may be sometimes chosen because it forms an azeotrope (e.g. benzene or the chlorinated hydrocarbons): the drying process can then be carried out simultaneously with the removal of solvent by distillation provided that (a) the boiling point of the solute is considerably higher than that of the solvent, so that when the latter is distilled off, there is no possibility of the dissolved substance being carried over, and (b) the solute is unharmed by being heated in the presence of water to the boiling point of the azeotrope it forms with the solvent. If these conditions cannot be met, or if, for other reasons, a solvent has been used which does not form an azeotrope with water, a desiccant must be used.

The use of a drying agent always introduces the risk of loss of solution (and hence, of solute) by absorption or adsorption on the desiccant; it is therefore important to use the smallest practicable amount of desiccant so as to minimize these losses. The choice of drying agent is often specified but the quantity required and the technique of its use is commonly left to the student's discretion, and it is essential to learn and practise the proper technique.

If the drying agent is not specified, it should be chosen on the basis of treating like with like: if in doubt, anhydrous magnesium sulphate or sodium sulphate (preferably the former) should be used, because there is then virtually no risk of incompatibility. Acidic drying agents must never be used with basic solutions, or vice versa. Potassium hydroxide (or sodium hydroxide) is very useful for drying basic substances: calcium chloride may be used for drying some neutral or acidic materials, but is incompatible with bases (such as ammonia or amines) or with alcohols, carbonyl compounds such as aldehydes.

Magnesium sulphate and sodium sulphate. These two desiccants are virtually interchangeable, but the magnesium compound is rather more rapid in its action. In their dried form they are fine, white powders capable of acting as chemically inert drying agents by absorbing water to form crystalline hydrates. The salts should be freshly dried before use by heating to constant weight in an oven at about 100–110°C: in any case, the dried material should be kept in well-stoppered bottles. To dry an ether solution, for example, the powder should be added in portions of about 1 g, and after each addition the mixture is shaken briskly. While water remains in the solution the added powder almost immediately absorbs it, and forms the crystalline hydrate, which loses its free-running powdery texture and tends to become translucent and dense. The powder is added until it no longer acquires this rather "heavy" texture and, on shaking, immediately and easily disperses, retaining the fluffy form, characteristic of the dry salt. If magnesium sulphate has been used, the drying action is almost instantaneous. The suspension of the sodium salt should be left for about an hour (preferably in a refrigerator) after the final addition, and shaken periodically; more powder should then be added if necessary. It should always be remembered that as these desiccants depend for their action on the formation of hydrated salts, which is a rapid process, once a stage is reached where the last portion added does not change its appearance and physical form (density), the drying can be assumed to be complete, and there is no point in adding more of the desiccant.

Calcium chloride. This is incompatible with some classes of organic compounds including amines, alcohols and some acids.

In many cases, drying with calcium chloride is required after a solvent extraction has been carried out in a separating funnel, and at least the first stage of the dehydration can be done without transferring the solution to another container. One single granule of the fused calcium chloride is crushed on a watch glass, or between the finger and thumb, and the powder sprinkled into the solution in the funnel, which is then gently swirled for about a minute. The solid will probably dissolve in the water present to give a dense liquid layer, which will quickly and cleanly settle to the bottom whence periodically it can be run off through the tap. The procedure is repeated until the solid fails to dissolve, although the surface mav become slightly sticky. The funnel is then set aside for about 30 minutes, and swirled occasionally; more calcium chloride may be added if necessary, but once the stage is reached where a small quantity of calcium chloride remains undissolved, dehydration is complete and nothing is to be gained by adding more. Accumulated calcium chloride solution in the bottom of the funnel may be run out through the stop-cock, but the dried organic layer should not be removed in this way because it would probably become contaminated with aqueous material; this dry solution should be poured from the neck through a small fluted filter paper to retain any traces of solid that may be present. The funnel is rinsed with a small quantity of dry solvent, the washings filtered and added to the main filtrate.

A minor variant of this technique is applicable if the solution to be dried is contained in a flask. The powdered calcium chloride is added portion-wise, as before, mixing well after each addition. The small amount of calcium chloride solution that accumulates initially is ignored and powder added until a small amount remains as a solid even after standing for 30 minutes. The dried solution is decanted through a fluted paper, and the flask rinsed with dry solvent as before.

Potassium hydroxide. Potassium hydroxide, preferably in the pellet form, is usually used to dry basic substances, such as amines, either alone or in non-aqueous solutions. A few pellets are added and the solution left with occasional swirling for at least 30 minutes. The pellets present a relatively very small surface area to the solution, in contrast with that provided by the powdered calcium chloride, for example, and to offset this limitation, which would normally result in very slow drying, it is permissible to add a considerable surplus; but it should never be necessary to use more than is required to give a layer one pellet thick in the bottom of the flask. It is advantageous to carry out the drying procedure on the liquid contained in a separating funnel, so that accumulated aqueous potassium hydroxide solution may be run off periodically, using a technique similar to that described for calcium chloride. Sodium hydroxide may be used instead of the potassium hydroxide, but the latter is usually to be preferred because it more rapidly absorbs water from the solution to be dried.

Drying Gases

This is a special application of separation in so far as the wet gas may be regarded as a mixture of two components, one of which is water vapour.

Gases are usually dried by contact with desiccants such as phosphorus pentoxide, silica gel, sulphuric acid, calcium oxide, calcium chloride or potassium hydroxide, but water may also be removed by freezing it out. This is done when using traps for high vacuum work.

The desiccant must be compatible with all the components of the gas mixture. For example, wet ammonia cannot be dried with concentrated sulphuric acid. Silica gel is chemically inert and can be regarded as compatible with all the commonly used gases. The oxidizing properties of concentrated sulphuric acid sometimes limit its usefulness as a desiccant: it cannot be used to dry hydrogen bromide. Syrupy phosphoric acid is a useful non-oxidizing alternative to sulphuric acid.

A quantity of desiccant cannot absorb an unlimited amount of water, and will eventually cease to function. Most desiccant materials cause appreciable resistance to gas flow, and with some, such as phosphorus pentoxide, this increases enormously as the drying agent changes its physical form on absorbing water.

In some desiccants, water saturation becomes self-evident. Thus, phosphorus pentoxide is converted from a white powdery solid to a sticky viscous liquid (metaphosphoric acid) and while the desiccant remains solid, it can be assumed to be still capable of absorbing water. Similar considerations apply to solid potassium hydroxide, sodium hydroxide or calcium chloride. Silica gel is commonly used for drying gases, largely because of its wide compatibility, but water absorption does not bring about any change in the physical form or texture. The degree of saturation of the silica gel with water could be determined by weight change, but this is laborious and it is convenient to use a modified form containing a cobalt salt; this imparts a blue colour when the gel is still capable of absorbing water, but hydration is accompanied by a colour change from blue to pink.

The following technique is recommended when filling a drying tube with phosphorus pentoxide, to avoid possibility of obstruction as the water is absorbed. If a "U" tube is used, a small, loose plug of glass wool is pushed down to the bottom, and then some of the phosphorus pentoxide powder is sprinkled on to give a heavy dusting. More glass wool is packed loosely into the tube and this again is dusted with the oxide. This process of alternately packing in glass wool and dusting with phosphorus pentoxide is continued until both limbs of the "U" tube are about three-quarters filled. The remaining space in each limb of the tube is packed with glass wool, without any phosphorus pentoxide. If a straight drying tube is used, the packing method is similar, except that it provides for each end being packed with a glass wool plug free from phosphorus pentoxide. The use of glass wool plugs to act as filters at the beginning and end of the filling operation is widely used when packing a tube with a solid desiccating agent.

Concentrated sulphuric acid is a very effective gas-drying agent, and is commonly used where chemical compatibility permits. The gas is usually bubbled through the acid in a suitable container, such as a Drechsel bottle (Fig. 6.1). Because of the highly corrosive properties of sulphuric acid it is especially important to prevent any possibility of it contaminating a chemical reaction mixture;

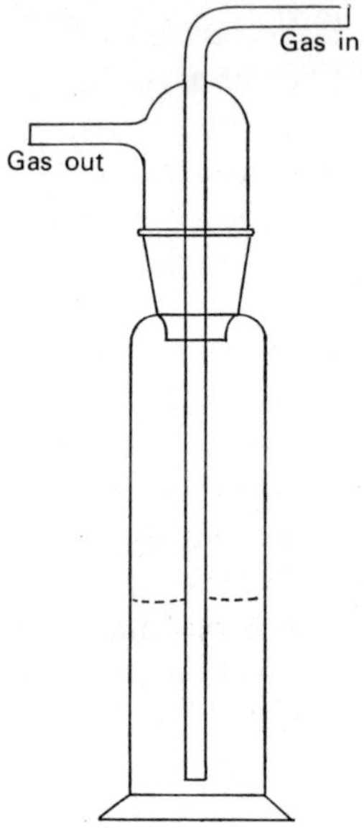

FIG. 6.1. Drechsel bottle.

it is desirable to provide empty traps before and after the sulphuric acid container, to accommodate any acid displaced by development of excessive positive or negative pressure. One of the big advantages of a liquid drying agent, such as sulphuric acid, is that the bubbling provides a constant visual indication that the gas is flowing. When using a dry desiccant, a bubbler system should be interposed between the drying train and the gas supply. It is desirable also to guard against the possibility of blockage developing somewhere in the system, either in the drying train itself or in the reaction mixture. To prevent possibly dangerous build-up of pressure under these conditions, the bubbler device should also be fitted with a liquid sealed blow-off tube (Fig. 5.14). This unit should be placed between the cylinder and the drying train. It is essential to prevent any suck-back of liquid into a gas cylinder, and an empty trap should be fitted between the cylinder and the first liquid container in the system. This trap should be large enough to accommodate all the liquid that can possibly be sucked back towards the cylinder: this precaution is particularly necessary when handling very soluble gases such as ammonia or hydrogen chloride.

All of the gas-drying systems so far discussed will inevitably cause appreciable resistance to gas flow. This is particularly objectionable when working with high vacuum systems, and therefore any kind of packed drying train must be avoided. Traps for such purposes are always based on the use of cooled surfaces to remove all condensable vapours. Typical glassware units for this purpose are illustrated in Fig. 5.11(a) and (b). The cooling media conventionally used for this purpose are either solid carbon dioxide with acetone or a similar solvent, or, for more exacting cases, liquid nitrogen. These traps can easily be designed to provide virtually no resistance to the gas flow, but accumulation of condensed materials must be removed before it can cause serious obstruction.

CHAPTER 7

CRITERIA OF PURITY

ALL the foregoing discussion about the preparation and purification of organic compounds has as its basis the characteristic properties of the covalent molecules that are encountered in this branch of chemistry. For a variety of reasons, a reaction intended to produce a required organic substance is likely to give a mixture possibly with many components, and much of the skill required is devoted to isolating the desired product in a suitable state of purity. Sometimes the purification procedures can be very tedious and exacting, because of the relative similarity of chemical and physical properties of the desired substance and its contaminants. It follows from this that there is a need for dependable methods for assessing the purity of organic substances, and a wide range of physical tests can be applied.

The most fundamental of these tests of purity is spectroscopic examination which may be carried out using radiation of various characteristic wavelengths. The traditional spectroscopic procedures utilize infrared, visible or ultraviolet radiation; more recent techniques include nuclear magnetic resonance.

Although spectroscopic measurements for the assessment of purity of an organic substance are of great importance, they are too advanced to fall within the scope of this book, and further reading is recommended.

The main requirements of a convenient and satisfactory test for determining the purity of an organic substance are as follows:

(a) The test should require only a small quantity of sample.
(b) The results should be obtained quickly and without great difficulty.

(c) The results should be reproducible and easily interpreted.

(d) The method of test should be as sensitive as possible, so that it will reveal very small amounts of contamination in an otherwise pure product.

The following methods of assessing purity will now be considered:

Melting Point
Refractive Index
Thin-layer Chromatography.

Determination of a Melting Point

A pure organic compound has a melting point with two essential characteristics; it is constant and it is sharp. In most cases the temperature range from the first indication of melting to the point where melting is complete should not exceed 2°C. In comparison with a pure substance, impure material melts at a lower temperature and over a much wider range. This phenomenon provides a very simple means of establishing the purity or identity of a substance, and it is common practice to recrystallize it until a constant melting point is obtained: each recrystallization operation removes some impurity and so tends to raise (and sharpen) the melting point, until the substance is virtually pure.

For elementary use, there are two satisfactory types of apparatus for determining melting points, the oil bath and the heated metal block. Both require a sample prepared in the same way.

The sample of solid, which must be dry, is crushed to a fine powder on a watch glass or microscope slide, using a spatula. The powder is introduced into a length of thin-walled glass tubing about 1 mm bore and 6 cm long. This tube may be drawn out from quill tubing, or supplies may be purchased ready made. One end of the tube is sealed by carefully rotating it in a microburner flame, and the sample introduced by gently pressing the open end into a small heap of the powder. The sealed end of the tube is tapped gently on the bench, to induce the powder to fall to the bottom. The process may be repeated as necessary to give a loose layer in the tube no more than about 3 mm in depth: it is important not to use too much sample.

If an oil bath is used for heating, the sample tube is attached to the thermometer by surface tension effects of a film of the oil. The thermometer bulb is dipped in the oil, and removed; then the

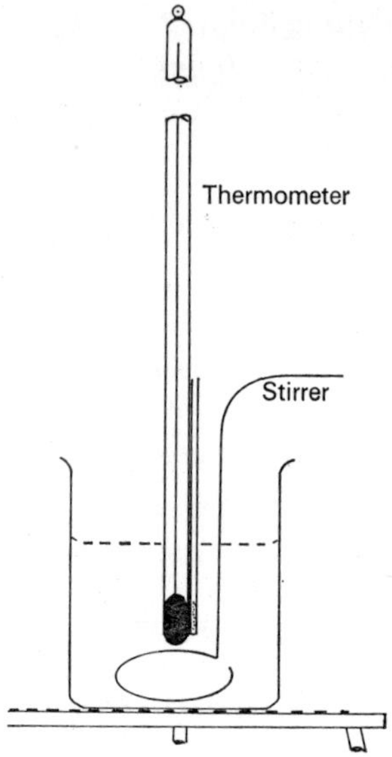

FIG. 7.1. Melting point apparatus.

tube is carefully attached, simply by pressing it into the oil film, so that the sample of powder is just about mid-way along the length of the thermometer bulb (Fig. 7.1). The thermometer with the attached sample tube is then gently lowered back into the oil bath. No further attachment of the tube is needed; if it tends to fall away from the thermometer, a new sample should be prepared using a longer capillary tube—say 8 cm long.

Stocks of capillary tubes for melting-point determinations should be stored in a desiccator, or at least in a sealed container to minimize the probability of deposition of a film of atmospheric moisture on the inner surface; this moisture could cause some inaccuracies in the melting-point determination.

The practical achievement of dependable, reproducible determination of melting points requires considerable care, and appreciation of the principles of heat transfer and temperature measurement.

Consider what happens when a melting point is determined by the usual elementary method. The sample, contained in a capillary tube, is attached to a thermometer, immersed in oil contained in a glass vessel heated by a small burner. Heat from the burner passes through the glass container to the oil, then to the thermometer bulb, which has to respond to the temperature increase. Simultaneously, heat from the oil passes through the walls of the glass capillary containing the crystalline sample; heat absorption there produces a rise in temperature until the solid starts to melt, but thereafter the temperature should remain steady because further heat input provides the latent heat of fusion. Only when all the substance is melted should the temperature rise further. It is essential that the rate of heat input during a melting-point determination should be slow enough to preserve thermal equilibrium at all times and to give the thermometer and the sample time to respond properly to the prevailing rise of temperature.

A satisfactory determination of melting point therefore requires the following rules to be rigorously observed:

(1) The sample should be powdered and absolutely dry before putting into the capillary tube. The amount used should be minimal; provided the sample can be clearly seen, it is sufficient. Usually it is enough to have about 2–3 mm depth in the tube, and if a magnifying glass is available, even less may be sufficient. The smaller the amount, and hence the heat capacity, of the sample, the more rapidly it responds to changes of temperature in the oil.

(2) The sample must be placed as near as possible to the bulb of the thermometer; the temperature it records must be that to which the sample is exposed.

(3) The bulb of the thermometer must be as small as possible, so that it responds rapidly to temperature changes in the heating medium.

(4) The rate of heating must be slow: when approaching the melting point, the rate of temperature rise must not be more than 5° per minute; a rate of 2° per minute is preferable. There are two main reasons for using a very slow rate of heating: the sample and the thermometer have different heat capacities but both must respond in unison as the temperature increases; the rate of heat input must therefore be very slow in order to avoid any possibility of one responding before the other. When the first sign of melting is detected, the temperature of the oil must be noted; this necessitates looking away from the sample to the scale of the thermometer, noting the position of the mercury thread, and observing the corresponding scale reading. Thus, merely noting the temperature takes a finite time, and it is obviously essential that during that period there should be no perceptible movement of the mercury.

(5) Excessive temperature gradients and erratic convection currents in the oil bath must be avoided, to ensure a steady heat input to the thermometer and sample. The oil should be stirred and protected from draughts.

(6) The thermometer must be calibrated. Most thermometers are marked to indicate the depth of immersion required to provide the declared accuracy. The stem correction is not usually very great—usually about one or two degrees in the region of 150°C—but this should be remembered when trying to account for discrepancies between observed melting points and those recorded in the literature. If this stem correction is allowed for in recording the melting point, the figure is said to be "corrected" and is usually reported as "m.p.   °C (corr.)".

When a melting point is determined, there is usually a temperature increase of 1–2° between the point at which there is the first perceptible change in the sample, to that at which the last trace of solid is converted to liquid; this is referred to as the "range" or "sharpness" of the melting point. When using melting point as a criterion of purity it is essential to take account of the range; it is not sufficient, and can be very misleading, to consider only one temperature. Differences of two or three degrees can easily arise when the melting point of the same substance is determined on different occasions because of such factors as stem correction, rate of heating, or simply inaccuracy of cheap thermometers. Thus, it is quite impossible to read any significance into melting points of two samples of the same substance having melting point quoted, for example, as 143°C for one and 145°C for the other. On the basis that impurities in a substance will depress the melting point, it might be assumed that the higher figure indicates a greater degree of purity, but if the respective melting points were fully quoted, as 142–143°C and 141–144°C, the first sample is revealed as being rather better than the second.

Mixed Melting Points: Identifying a Compound

Very many different substances may have the same melting point, so this alone is not sufficient evidence to establish the identity of a particular compound. Confirmation may, however, be obtained by a mixed melting point determination. Suppose two substances *A* and *B* have the same melting point—say 161–2°C. In order to determine if *A* and *B* are identical or different, a mixture of the two in approximately equal proportions is prepared, and the melting point determined. If *A* and *B* are identical, the melting point will still be 161–2°C, but if they are not of the same chemical composition, the value will be considerably below 161°C and also much less sharp; typical figures would be 152–159°C.

For the determination of mixed melting points, two components should be very well mixed in approximately equal proportions, by crushing them together with a spatula on a watch-glass or a microscope slide; to avoid errors arising from different rates of heating three samples, each in its own melting-point tube, of *A* alone, *B* alone and the mixture of *A* and *B*, should be attached to the thermometer, and the melting point of all three determined simultaneously.

Refractive Index

Refractive index measurements on materials that are liquid at ordinary temperatures can provide a very rapid and sensitive means of determining purity.

For most student work the refractive index of a sample is measured on an Abbé refractometer; only about two small drops of sample are required for this measurement because the instrument merely requires the formation of a continuous film of liquid between the accurately ground faces of two prisms, which are brought into close proximity. The refractometer is usually calibrated to provide readings of the refractive index, calculated with reference to the wavelength of the sodium D line, but the instrument itself requires only white light, supplied either by daylight reflected from a mirror or by a small lamp. Even a simple type of instrument provides readings of refractive index to the fourth place of decimals. Refractive index is temperature dependent: as a guiding rule, it can be assumed that for each 1°C rise in temperature, the refractive index decreases, by 0·0004. All refractive index measurements are therefore made and compared with due regard to the temperature; the sample holder of most commercially available refractometers is provided with a jacket through which water from a thermostatically controlled source can be circulated. In some cases it is not necessary to control the temperature, but only to know its value, but if the refractive index of a volatile sample is being measured, especially in hot weather, there is a possibility of evaporation being so rapid that the sample is lost before the measurements can be completed. In this case it is necessary to cool the jacket of the refractometer, or to use a larger quantity of sample.

Usually literature reports of the refractive index of a substance are related to the sodium D line at a specific temperature, usually between 15–20°C. The method of recording the refractive index is exemplified as follows: $n_D^{18} = 1{\cdot}3721$. The subscript D refers to the sodium D line, and the superscript to the temperature.

As the detailed design and method of operation of individual types of commercially available refractometer differ, detailed instruction for use cannot be given here. However, in one very commonly used type, the sample is placed upon the surface of a prism, made of relatively soft glass, easily damaged by scratching or chipping. The application of the droplet sample to this glass surface should be made with considerable care; glass rod is too hard, and may damage the surface. It is desirable to use a small dropping-rod made of an inert plastic material, such as polyethylene, but a very small pipette may be useful in some cases.

The refractometer must be cleaned thoroughly immediately after use. For most purposes ether is a suitable cleaning solvent. A small swab of cotton wool or fine paper tissue is soaked in ether, and used to wash the whole area of the two faces of the prism and its holder coming into contact with the liquid sample during a refractive index measurement. The washing operation is repeated twice, using a clean swab each time. Finally the washed faces of the prism should be dried with a dry piece of clean cotton wool or paper tissue. Solvents other than ether may be used if necessary.

INTERPRETATION OF REFRACTIVE INDEX MEASUREMENTS

Because of the temperature dependence of refractive index, comparison of the values obtained can only be made if the figures all relate to the same temperature. Even if corrections are made for this, refractive index is so sensitive to temperature (0·0004 per °C) and also to quite small amounts of contamination that little significance need be attached to variations of 4 units or less in the fourth place of decimals. As a general rule, any discrepancy exceeding about 0·0007 is likely to indicate significant difference in composition.

It is sometimes convenient to use refractive index measurements as a means of determining the approximate composition of a mixture of which the nature of the constituents is known. In many cases of mixtures of closely similar chemical composition, the refractive index will be approximately a linear function of molar composition. For example, to obtain a graph of refractive index against composition for a mixture of butyl chloride and butyl bromide, the refractive index for 100 molar per cent butyl chloride and the corresponding figure for 100 molar per cent butyl bromide are plotted and joined by a straight line. Using this graph, it is possible to use observed refractive index measurements obtained from mixtures of these two halides to estimate the composition. The method is not highly precise, but can be quite useful for following the course of a reaction or a fractional distillation.

Chromatography

This technique can be applied very effectively either for analytical purposes or for carrying out separations on a multicomponent mixture as part of a preparative procedure. The technique of thin-layer chromatography (TLC) provides a very elegant and simple analytical tool for demonstrating the purity (chemical homogeneity) or otherwise of an organic substance. When a solution of an organic substance comes into contact with a solid, such as kieselguhr or silica gel, the solute molecules are adsorbed on to the surface. In very elementary terms, the dissolved organic material can be considered to be attracted to the solid surface, and this attractive force—the strength of adsorption—is a function *inter alia* of the nature of the inorganic adsorbing surface and of the structure of the organic molecules. If the nature of the absorptive surface and the solvent system are constant, small differences in molecular structure can have a considerable effect on the strength of adsorption, and hence on the ease with which the molecule can be desorbed—that is, detached from the adsorptive surface by the movement over it, of another solvent. This is the property utilized in TLC. The adsorptive inorganic solid (kieselguhr or silica gel) is deposited on a clean glass plate to form a smooth even layer, usually about 100μ ($0 \cdot 1$ mm) thick. Consider a mixture containing three organic components, *A*, *B* and *C*, dissolved in a solvent such as ethyl acetate. If one drop of this solution is deposited on the coating of the plate, the organic molecules will be adsorbed and the solvent, which is chosen to be much less strongly absorbed, will diffuse through the inorganic layer and quickly evaporate. The edge of the plate is now dipped into a trough of a suitable solvent, so that it can gradually diffuse up through the coating. When the diffusing solvent encounters the deposit of *A*, *B* and *C*, an equilibrium will be established between the two opposing forces—the absorptive force tending to "stick" the molecules on to the inorganic solid, and the desorbing forces tending to permit the organic material to be carried along with the moving stream of solvent. The only significant variable in this system is the different absorptive forces between the coating surface and the different molecular types, *A*, *B* and *C*. As the solvent moves past the spot containing the three components, the least strongly absorbed (say *A*) will tend to become detached from its original position on the plate, and be carried along with the moving solvent. Thus, the action of the moving solvent is to spread the three components *A*, *B* and *C*, which, as a result of the differences of equilibrium between the adsorptive and desorptive forces, travel along in the direction of the moving solvent at characteristic rates which, other things being constant, are a function of their molecular structures. For each solvent system and each type of adsorbent, pure organic substances will have a characteristic R_f value, which is defined as:

$$R_f = \frac{\text{Distance moved by substance}}{\text{Distance moved by solvent front}}.$$

Because the TLC method is so very sensitive, it is possible to separate and even identify the components of a mixture of closely similar substances by this technique. Conversely, if a substance suitably treated by this technique shows no signs of separation, it is very pure indeed—chromatographically pure.

The separate components may not be directly visible, and there are several techniques for exposing the plate to some other material, which will result in the formation of a highly coloured complex, easily visible at the site of the spot. Iodine vapour is used in some cases. Amino acids are made visible on the plate by converting them into their coloured complexes formed with ninhydrin. In some cases again, exposure of the plate to ultraviolet light makes the organic spots visible by fluorescence.

The separation of ortho- and para-isomers of nitroacetanilide will serve to illustrate the essential techniques of TLC.

PREPARATION OF THE PLATE

Students will not be required to coat their own glass plates. It is assumed here that supplies of ready-made plates are available.

It is necessary for the solvent to move forward across the plate so that the front edge—the solvent front—remains straight and perpendicular to the direction of movement; to encourage this, the solid on the long edges of the coated plate is removed, because the layer here may be of uneven thickness and hence result in solvent diffusing at a rate different from that across the rest of the coating. This surplus solid is removed by rubbing the ball of the thumb along those two edges, in order to wipe off a strip of the coating about 3 mm wide.

Solution for Spotting

The sample to be examined chromatographically is dissolved in any suitable solvent, such as ether or ethyl acetate; the solution should preferably be saturated.

SPOTTING

It is common practice to put about two or more different samples for examination side by side on the same plate. For many purposes a plate about 8 × 2 in. is useful. The plate is placed on the bench, with the coated surface uppermost, and the spots (see later) are deposited in a straight line, about 1 in. from one of the short edges. It is not permissible to mark a line on the coating, but the location can be indicated by placing a straight-edge over the position required, taking care that this marker does not touch or damage the coating. The position of this imaginary line may be marked by a small nick made on each edge of the coating.

The spot is deposited from a very fine dropping pipette, prepared from a piece of tubing drawn out to give a capillary about 0·3 mm bore. A piece of this capillary, open at both ends, is dipped into the solution to be spotted, which rises up the bore, and hangs as a very fine droplet from the bottom. This droplet is very carefully deposited on to the coating on the glass plate, without allowing the tube itself to touch or perforate the coating. This is a very delicate operation, in which the right hand carrying the capillary should be supported by the left resting on the bench. The tube is very cautiously lowered until the hanging droplet just touches the coating to give a small circle of liquid.

Each spot should be as even and as round as possible, and not more than about 2–3 mm in diameter. If the solution being spotted is not very concentrated, the one small drop may not deposit sufficient solute. In that case a heavier deposit is obtained, not by putting on a larger drop, but by adding a succession of drops, one on top of the other and allowing the solvent to evaporate from each addition before the next one is made. In this way, a large concentration of the solute is deposited on the plate in the same small round area of spot. If more than one sample is being spotted on to the plate, they are deposited along the same line (indicated but not acutally marked) so that each one is at least 1 in. from its neighbour.

Having completed the spotting, the solvent is allowed to evaporate completely.

DEVELOPMENT

This is the process whereby the selected solvent—ethyl acetate will serve in the example under discussion—is made to migrate up the plate.

For this purpose, the solvent is contained in a tank to form a layer about 1 cm deep. The distribution of the solvent vapour in the tank should be as even as possible, and in particular, it is important to prevent solvent evaporating from the coated surface of the plate while development is in progress. To meet this need the development tank is provided with a large sheet of filter paper, folded to form a three-sided, box-like section, which is stood on its edge, dipping into the layer of solvent. In this way, the solvent rises by capillary up the paper, and evaporation from the surface provides a means of producing even and maximum vapour distribution. The tank is provided with a well-fitting lid, usually consisting of a glass plate with ground edges.

The coated chromatographic plate to be developed is placed in the tank in such a way that the shorter edge is immersed in the solvent deeply enough to wet the first half centimetre or so of the coating, but the line of spots on the plate must be well above (about 1 cm) the liquid level. The plate is supported in a near-vertical position, and with the lid on the tank, solvent is allowed to rise by capillarity through the surface of the coating. Soon after starting to rise, the liquid front goes over the line where the spots were deposited, and the ascent is then allowed to continue until the solvent front (the moving edge of the rising liquid) reaches to within about 1 in. of the top of the coated area.

The plate is then removed from the tank and the position of the solvent front on the coating marked by scratching a line with a pin across the coating.

The plate is left to stand in air until the solvent has completely evaporated from it. In the case of ethyl acetate this can easily be judged by smell.

The spots on the plate are now almost invisible, but they can be located very easily by exposure of the plate to iodine vapour. This treatment is carried out in a tank similar to that used for the development, but containing a sprinkling of iodine crystals on the bottom. The plate is placed in the tank, again in a near-vertical position, with the origin line at the lower end. The lid of the tank is replaced, and the plate left exposed to the iodine vapour. After a very short time the position of the spots becomes apparent by the formation of a yellow complex between the iodine and the compounds on the plate. The iodine exposure is allowed to continue long enough for all the spots to be clearly visible. The plate is then removed from the tank.

Since the iodine complex is not stable in air, the colour fades and the position of all the spots should be marked by using a fine pin to scratch through the coating around the circumference of each one. An illustration of the typical results obtained is given in Fig. 7.2.

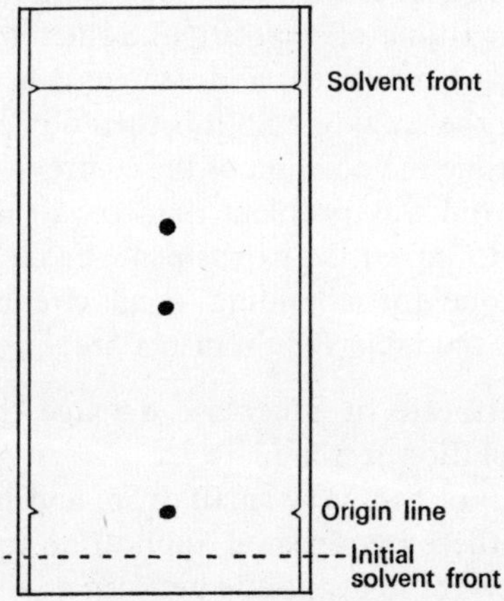

FIG. 7.2. Thin layer chromatography plate.

Even a small amount of contamination of one isomer with the other can easily be detected by the TLC technique, and this therefore provides an excellent means of assessing purity. If, on submitting a sample to thin-layer chromatography, only one spot is detected, the compound is demonstrated to be pure and free from contamination with the other isomers.

Measurement of R_f Values

As has already been shown, the R_f value is defined as

$$\frac{\text{Distance moved by spot}}{\text{Distance moved by solvent}}.$$

The solvent front usually moves forward in a straight line parallel to that along which the spots were originally deposited; it is therefore not difficult to measure the distance between the original line where the spots were deposited and the position finally reached by the solvent. It is not always so easy to determine the distance moved by the spot itself. If the chromatography has been done well, the moving spot will retain its small circular form. By convention the distance is then measured from the centre of this, back to the original line of deposition. Sometimes, however, the spot becomes grossly enlarged, and deformed in shape during its progress through the coating, and it is then obviously much more difficult to determine the position of the centre with any precision. It is better to avoid this problem than try to deal with it, and usually it is best to repeat the experiment. From the viewpoint of operating technique for obtaining small circular spots on the developed phase, the main requirements are:

(a) Do not perforate or otherwise damage the coating when applying solution to the plate.

(b) Apply the spot as a very small drop, and build up the concentration there by repeated application to the same place, allowing solvent to evaporate between additions. In this way spreading of the applied solution over a large area of the coating is avoided.

CHAPTER 8

PACKING AND STORING PRODUCTS AND SAMPLES

Solids

Glass containers are suitable in all cases. Small tubes made of polyethylene or polypropylene are also available, but over long periods of time there is a possibility of absorption of traces of the product by the plastic of which the container is made. For this reason it is desirable not to use a plastic container that has already been used to store some other organic material. The container should be tared before filling, and the gross, tare and net weights recorded on the label.

Small quantities of solids should be tipped on to a piece of clean, glazed paper, which has been folded down the middle to give a V shape. With this paper chute the solid can easily be transferred to the container. To avoid loss by spillage, the tube should be stood on a large piece of glazed paper on the bench, so that any solid falling there during the transfer can easily be recovered in a clean condition. Filter paper should never be used for this operation, because fibres from it may contaminate the sample.

If a cork is used to close a sample container, it should be covered with cellophane, greaseproof paper, or aluminium foil, so that the contents do not come into contact with the cork itself. The cork should be wrapped completely in the paper, which is secured by twisting (Fig. 8.1); the paper-covered closure is then inserted firmly into the mouth of the glass container.

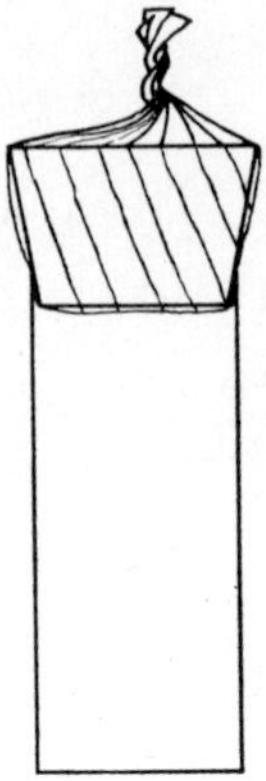

FIG. 8.1. Vial for solids.

Liquids

Packing liquids for permanent storage presents rather more problems. It may sometimes be sufficient to keep the liquid in a specimen tube fitted with a cork or rubber bung, which should be protected with a film or foil wrapping as described above. Larger quantities of liquids are best kept in bottles—preferably of brown glass to exclude light which may be harmful—closed with a glass stopper, cork, bung or a plastic screw cap.

Small samples are best stored in sealed ampoules, i.e. glass tubes closed by sealing the glass in a flame. In this way the contents are in contact only with glass, and if the seal is properly made there is no possibility of leakage or ingress of air. Ready-made ampoules are available in a range of sizes from about 0·5 ml upwards, but they can also be made quite easily in the laboratory from a small test tube, or other glass tubing. The aim is to draw out in the glass tube a constriction so positioned that the volume between the original base of the tube and this newly produced neck is about twice that of the liquid to be accommodated. To facilitate manipulation a temporary handle is fixed to the original lip of the test tube by sealing on a piece of glass rod. The tube is then drawn out, after heating it in a blow-pipe flame. The constriction should be positioned to allow at least 2 cm of the original tubing above it to serve as a built-in funnel. The internal diameter of the neck should be about 2–3 mm at its narrowest point (Fig. 8.2).

Fig. 8.2. Ampoule for **liquids.**

FILLING THE AMPOULE

One method of filling is to place some of the liquid into the mouth A of the ampoule prepared as described above, and then to warm and cool B alternately by immersion in hot andc old water. In this way air is expelled from B and replaced by the liquid in A. Care must be taken to avoid having too much liquid in A when air is being expelled from B. Failure to take this precaution can result in loss of liquid by sputtering as the air bubbles through. It is also essential to avoid over-filling B. The level of liquid should be well below the shoulder of the ampoule; it is usually desirable to leave at least 2 cm between the liquid level and the point at which the seal is to be made, so that the heating involved in this does not damage the contents. The internal surface of the neck must be thoroughly cleaned to remove all organic material left there in the process of filling, by wiping with a small spill of filter paper. Cotton wool should not be used because it may deposit fibres. If organic material is left on the inner surface of the neck, heating to produce the seal will result in charring and an unsatisfactory closure.

Another method of filling is to use a separate small funnel, usually made from a test tube drawn out to provide a long slender stem that will pass easily through the neck of the ampoule and reach well below the constriction. After filling, if the funnel is carefully withdrawn, no cleaning of the neck of the ampoule should be necessary.

The ampoule is sealed by rotating in a blow-lamp flame, placed so that it just heats the narrowest part of the neck. During this heating the ampoule is supported and rotated between finger and thumb. The neck should not be drawn out immediately the glass has softened, because this will result in a very fragile, hair-like capillary seal: when the glass has softened, it is kept heating and rotating while it collapses inwards on itself until the bore has closed. Rapid drawing-out will then permit removal of the surplus glass, leaving a strong well-rounded sealed tip on the neck of the ampoule.

Small quantities of liquids can be kept in test tubes for short periods. Corks or rubber bungs wrapped in foil or paper should be used as stoppers. The use of standard taper ground-glass equipment for storage of liquids is not to be encouraged. Although the liquid will not deteriorate, the joints are apt to seize on prolonged standing, and it may be rather difficult to separate them. Also this type of glassware is expensive, and should not be put out of general laboratory use in this way.

Labels for Samples, etc.

It is commonly necessary to store for future use, materials prepared in the course of work in the laboratory. These samples must be properly labelled to permit their complete identification at any time. The minimum information recorded on the label should be:

(a) name of the substance,
(b) reference number or other identification,
(c) physical constants, e.g. melting point, boiling point, refractive index,
(d) date of preparation,
(e) weight (tare) of container, so that the weight of the contents can rapidly be determined.

A specimen label is shown below.

ACETANILIDE
$C_6H_5NH \cdot CO \cdot CH_3$

RECRYST. EX. 5% AQUEOUS ACETIC ACID

m.p. 113–114°C

G 143 g	W.S. 123/a.
T 72 g	3.v.64.
N 71 g	

An unlabelled sample is valueless and care must be taken to ensure that the label is securely fixed to the container itself and not to the cork or other removable closure. Adhesive transparent cellulose tape should be used to protect the whole of the area of the label, overlapping sufficiently to assist in fixing the paper to the main surface of the container. It is not satisfactory to use tie-on labels except for temporary purposes.

CHAPTER 9

REPORTS AND COMMUNICATIONS

IT IS essential to keep adequate written records of all experimental work, indicating what was done, when it was done, and the results obtained. This record must be completed in the laboratory as soon as the necessary information is available and the natural tendency to procrastinate must be resisted.

The essential purpose of these records is to remind the operator himself of the detail of the experiment long after it was done, and —a more exacting need—to convey to another chemist enough information to permit him to repeat the experiment or to make it possible for him to evaluate accurately the significance of the results obtained.

The best test of the adequacy of the record is for the student to imagine he is being presented with the record in question, and to consider whether he could repeat the experiment satisfactorily using that information alone.

Although in this book great emphasis has been laid on the detailed techniques of carrying out the commonly used unit operations of organic chemistry, it is not normally necessary to record all this in a laboratory notebook once the operator has gained reasonable skill in these techniques. Thus, if a substance is being recrystallized, it is usually not necessary to record any more detailed information than the nature and quantity of the solvent and the weight of the solute before and after crystallization. The fact that the solvent was heated to boiling and then cooled adequately is inferred, unless there is a statement to the contrary.

An example of the way a preparation should be recorded is given on p. 193. The account should always begin with the equation for the reaction and the molecular weights of the reagents and products. The record must obviously show the weights (and/or volumes) of the reagents used, as well as the yield of product which should be reported as a weight and as a percentage of the theoretical amount.

All temperatures and times of heating should be noted, except that for solutions boiling under reflux, where the boiling point of the solvent is known and not subject to significant variation. In particular, a statement of the boiling points of liquids distilling under vacuum is valueless unless the pressure is also recorded. A record of a fractional distillation has already been shown on p. 96.

It is very important that the records should include any information that will help to assess the results of an experiment, especially if it is known that something went wrong with it. Thus, if a yield is given as "10·8 g (43 per cent of theory)", it is obviously necessary to consider what happened to the lost 57 per cent. Was this loss of 57 per cent of the theoretical yield due to the inherent chemical limitations of the method of preparation, or was it because some of the product was lost when it was dropped on the floor?

The importance of this is obvious if one is using such records to decide which method to use to carry out a reaction. A choice based on the simple statement that reaction *A* gave 40 per cent yield while reaction *B* gave 55 per cent will not be very satisfactory if the record omitted to mention that in carrying out reaction *A*, half the product was spilled before weighing it.

Apart from accidents, low yields may be due to:

(a) Incomplete conversion of the starting materials. If some of the starting materials have been recovered, then a distinction must be drawn between percentage yield calculated on the amount of reactants in the original mixture (i.e. the conversion) and the yield calculated on the basis of starting material actually consumed, after allowing for recovery.

(b) Side reactions resulting in consumption of starting materials to give products other than the desired one.

(c) Losses of reaction product during isolation and purification. The amount of the crude product at each stage should always be recorded so that these losses can be assessed.

There is much to be gained in setting out a "materials balance sheet", giving as much information as possible about the materials consumed and produced, under such headings as "Product", "Unreacted starting material recovered", "Solvent recovered", "By-products" (e.g. polymer remaining after a distillation).

A report of an experiment in a laboratory notebook should always include the date of starting it, and the dates on which subsequent operations were carried out. This is simply a matter of general good practice for students, but in professional work it can assume considerable economic importance if patents become involved. When an invention is made the subject of an application for patent protection, the date of the discovery may be vitally important if a dispute concerning which of two distinct applications for the same invention arises, and the award of patent will go to the "true and first inventor". A court ruling on this matter may depend ultimately on the date indicated in the laboratory notebook.

It is also necessary to assign a serial number or other method of identification to each experiment to facilitate subsequent reference to it.

On the general question of laboratory notebook records, the essential requirements are clarity and brevity. For student work, a terse, abbreviated form of writing, as indicated in the specimen, is best, provided that there is no sacrifice of accuracy and clarity. Since a laboratory notebook should provide a record of what was actually done in the laboratory, it should be written in the impersonal past tense. It is wrong to write these laboratory records either in the form of a first-person diary or, what is worse, as a series of instructions transcribed directly from a laboratory instruction sheet.

APPENDIX

SPECIMEN LABORATORY NOTEBOOK RECORDS

A. Ex. 1. 14.12.65

PREPARATION OF ETHYL BENZOATE FROM BENZOIC ACID

$$\underset{\substack{122\\25}}{C_6H_5COOH} + \underset{\substack{46\\9\cdot1}}{C_2H_5OH} \xrightarrow{H_2SO_4} \underset{\substack{150\\30}}{C_6H_5\cdot CO\cdot O\cdot C_2H_5} + \underset{\substack{18\\3\cdot6}}{H_2O}$$

90 ml (75 g) Ethanol in 500 ml R.B. flask—added 5·5 ml (10 g) conc. sulphuric acid cautiously with thorough mixing between each portion added—cooled flask under cold water tap to dissipate heat liberated—(maximum temp. 42°C). Added 25 g (0·2 mole) benzoic acid—fitted flask with Liebig condenser—heated reaction mixture under reflux on steam bath for 1½ hours. The benzoic acid dissolved within a few minutes of the reaction mixture starting to boil, which thereafter remained homogeneous, but gradually acquired a faint but distinct pink colour—apparatus then arranged for vacuum distillation (water pump) using the same reaction flask—distilled off excess ethanol as rapidly as possible (*ca.* 20 min) using steam bath—collected *ca.* 70 ml distillate. Residual liquid in boiler flask (*ca.* 35 ml) poured into water (150 ml) contained in 500 ml separating funnel—mixed well—allowed to settle and removed upper (non-aqueous) layer. The lower (aqueous) layer was extracted twice with ether (2 × 50 ml) and the ether extracts combined with the original non-aqueous layer—ether solution washed with (a) water, (b) 10 per cent Na_2CO_3 solution, (c) water (50 ml each wash)—dried over $CaCl_2$

—removed ether by distillation—ethyl benzoate then distilled (air condenser) and collected main fraction—a clear, bright, colourless liquid, b.p. 208–212°C—residue in boiler flask 1·5 g.

Yield	21·4g (see note 1)
	70·5% of theory
b.p.	208–212°C (lit. 212°)
n_D^{17}	1·5061 (lit. n_D^{17} 1·5068) (see note 2)

Notes

(1) This yield was rather lower than it should have been because of spillage of some of the ether extract (estimated to be about 10 per cent of the total volume). On this basis, the yield should have been about 23·5 g (77 per cent of theory).

(2) The refractive index is slightly low; this suggests some slight contamination—this was not identified, but was probably dissolved water.

1.A. 18.12.65

ETHYL BENZOATE—REPEAT PREPARATION

The above preparation was repeated, using the same quantities of reactants, to determine the yield when no accidental loss occurred.

Yield	23·4 g
	78·0% of theory
b.p.	208–212°C
n_D^{18}	1·5058

Ex. 2. 19.12.65

PREPARATION OF ACETANILIDE FROM ANILINE

$$\underset{93}{C_6H_5NH_2} + \underset{82}{CH_3COONa} + \underset{102}{(CH_3CO)_2O} + \underset{36}{HCl} \longrightarrow \underset{135}{C_6H_5NH \cdot COCH_3}$$

Aniline (9·3 g) (9·2 ml) (0·1 mole) dissolved in 40 ml 2N HCl in 500-ml conical flask—the resulting solution of aniline hydrochloride was swirled while adding rapidly and in one portion 12·1 g (11·2 ml) acetic anhydride (0·12 mole) followed immediately by 15 g sodium acetate trihydrate (0·11 mole) which was also added in one portion.

The reaction mixture was kept swirling for about 5 min during which time the temperature rose to about 38°C and crystals of acetanilide separated—it was then left to stand at room temperature for another hour. The crystals were filtered in a Buchner funnel (11 cm) pressed down well, and washed with cold water (10°C) (*ca.* 50 ml). After pressing the crystals well to remove as much water as possible, they were spread to dry on a large watch-glass in an oven at 60°C.

Yield crude acetanilide = 11·4 g (84·6% of theory)
m.p. = 109–113°C

The crude acetanilide was recryst. from boiling 10 per cent acetic acid (in water) (120 ml). The solution in the boiling aqueous acetic acid was filtered rapidly through a fluted filter paper (Whatman No. 54) in a glass funnel previously heated to *ca.* 100°C in a bunsen flame. The filtrate, almost colourless, and quite bright, was left standing undisturbed at room temperature until crystallization was well developed. The mixture was then cooled to 8°C in an ice–water bath. The crystals were filtered in a Buchner funnel, pressed down well, washed with cooled water (8°C) (3 × 25 ml) and finally dried in an oven at 60°C—product was white (colourless ?) needle crystals.

Yield of recrystallized product	= 10·1 g
m.p.	= 113–114°C
Return on recrystallization	= 88·6%
Overall yield from aniline	= 74·8% of theory.